PERSPECTIVES ON USER INNOVATION

Series on Technology Management*

Series Editor: J. Tidd (Univ. of Sussex, UK) ISSN 0219-9823

Published

Vol. 6   Social Interaction and Organisational Change
         Aston Perspectives on Innovation Networks
         edited by O. Jones *(Aston Univ., UK)*, S. Conway *(Aston Univ., UK)*
         & F. Steward *(Aston Univ., UK)*

Vol. 7   Innovation Management in the Knowledge Economy
         edited by B. Dankbaar *(Univ. of Nijmegen, The Netherlands)*

Vol. 8   Digital Innovation
         Innovation Processes in Virtual Clusters and Digital Regions
         edited by G. Passiante *(Univ. of Lecce, Italy)*, V. Elia *(Univ. of Lecce, Italy)* & T. Massari *(Univ. of Lecce, Italy)*

Vol. 9   Service Innovation
         Organisational Responses to Technological Opportunities and Market Imperatives
         edited by J. Tidd *(Univ. of Sussex, UK)* & F. M. Hull *(Fordham Univ., USA)*

Vol. 10 Open Source
         A Multidisciplinary Approach
         by M. Muffatto *(University of Padua, Italy)*

Vol. 11 Involving Customers in New Service Development
         edited by B. Edvardsson, A. Gustafsson, P. Kristensson,
         P. Magnusson & J. Matthing *(Karlstad University, Sweden)*

Vol. 12 Project-Based Organization in the Knowledge-Based Society
         by M. Kodama *(Nihon University, Japan)*

Vol. 13 Building Innovation Capability in Organizations
         An International Cross-Case Perspective
         by M. Terziovski *(University of Melbourne, Australia)*

Vol. 14 Innovation and Strategy of Online Games
         by Jong H Wi *(Chung-Ang University, South Korea)*

Vol. 15 Gaining Momentum
         Managing the Diffusion of Innovations
         edited by J. Tidd *(University of Sussex, UK)*

Vol. 16 Perspectives on User Innovation
         edited by S. Flowers & F. Henwood *(University of Brighton, UK)*

*For the complete list of titles in this series, please write to the Publisher.

SERIES ON TECHNOLOGY MANAGEMENT – VOL. 16

PERSPECTIVES ON USER INNOVATION

editors

Stephen Flowers
University of Brighton, UK

Flis Henwood
University of Brighton, UK

Imperial College Press

Published by

Imperial College Press
57 Shelton Street
Covent Garden
London WC2H 9HE

Distributed by

World Scientific Publishing Co. Pte. Ltd.
5 Toh Tuck Link, Singapore 596224
USA office: 27 Warren Street, Suite 401-402, Hackensack, NJ 07601
UK office: 57 Shelton Street, Covent Garden, London WC2H 9HE

British Library Cataloguing-in-Publication Data
A catalogue record for this book is available from the British Library.

PERSPECTIVES ON USER INNOVATION
Series on Technology Management — Vol. 16

Copyright © 2010 by Imperial College Press

All rights reserved. This book, or parts thereof, may not be reproduced in any form or by any means, electronic or mechanical, including photocopying, recording or any information storage and retrieval system now known or to be invented, without written permission from the Publisher.

For photocopying of material in this volume, please pay a copying fee through the Copyright Clearance Center, Inc., 222 Rosewood Drive, Danvers, MA 01923, USA. In this case permission to photocopy is not required from the publisher.

ISBN-13 978-1-84816-699-8
ISBN-10 1-84816-699-0

Typeset by Stallion Press
Email: enquiries@stallionpress.com

Printed in Singapore.

Contents

Chapter 1.	Introduction: Perspectives on User Innovation *Stephen Flowers and Flis Henwood*	1
Part I:	**Exploring the Role(s) of Users in Innovation**	**9**
Chapter 2.	The Historical Construction of User Innovation *Georgina Voss*	11
Chapter 3.	The Dynamics of User Innovation: Drivers and Impediments of Innovation Activities *Christina Raasch, Cornelius Herstatt and Phillip Lock*	35
Chapter 4.	Intermediaries, Users and Social Learning in Technological Innovation *James Stewart and Sampsa Hyysalo*	57
Part II.	**Drawing Users into the Innovation Process**	**89**
Chapter 5.	User-Centric Innovations in New Product Development — Systematic Identification of Lead Users Harnessing Interactive and Collaborative Online-Tools *Volker Bilgram, Alexander Brem and Kai-Ingo Voigt*	91
Chapter 6.	Proactive Involvement of Consumers in Innovation: Selecting Appropriate Techniques *Karen L. Janssen and Ben Dankbaar*	131
Chapter 7.	User-Producer Interactions in Emerging Pharmaceutical and Food Innovations *E. H. M. Moors, W. P. C. Boon, R. Nahuis and R. L. J. Vandeberg*	161

Part III.	**New Directions in User Innovation Research and Policy**	189
Chapter 8.	Outlaw Community Innovations *Celine Schulz and Stefan Wagner*	191
Chapter 9.	User Innovation: The Developing Policy Research Agenda *Stephen Flowers*	211
Chapter 10.	The Freedom-Fighters: How Incumbent Corporations are Attempting to Control User-Innovation *Viktor Braun and Cornelius Herstatt*	229
	Index	259

INTRODUCTION: PERSPECTIVES ON USER INNOVATION

STEPHEN FLOWERS and FLIS HENWOOD

User innovation is becoming more widely recognised as a potent force in many parts of the economy and society, with significant theoretical, practical and public policy implications. However, these implications can often be overlooked, as they often do not fit easily in the standard sectoral or functional discussions that tend to frame many of the debates concerning innovation. This edited collection is a contribution to the ongoing discussion over the part played by users in innovation, and to the wider debate concerning the shape and structure of the innovation process itself. The papers in this book reflect this debate and are contributions drawn from three potentially complementary research traditions: Innovation Studies, Science and Technology Studies and Innovation Management. User involvement in innovation is complex and multi-faceted phenomenon and a thorough understanding of this process can perhaps only be reached by drawing on different disciplinary approaches and perspectives. This collection brings together these different traditions and explores their contribution to understanding user innovation as an important theoretical and practical endeavour in its own right.

Innovation is often a challenging, complex, and contested process with users playing a central role in the creation, shaping and diffusion of new products, services and ideas. Users are clearly of great importance, but sometimes in potentially conflicting ways. Users can be highly active innovators in their own right, creating and distributing their own products or services. Some users may prefer to modify products that they have purchased or they may wish to provide feedback to firms so that they can improve the products they are able to buy. Alternatively, other users may decline any involvement with a product or service by passively resisting its diffusion. Indeed, some highly creative users can create products, services or workarounds that compete with or bypass mainstream offerings, thereby more actively resisting "official" innovations.

The academic understanding of the role of the user within the processes of innovation tends to be fragmented, with different strands of literature focusing on particular aspects or perspectives. Different strands of literature tend to be framed around a particular story or meta-narrative in which users are perceived as passive "customers", active "shapers" or useful "contributors" to innovation processes. Innovation processes themselves may be located within market-based relationships in which organisations seek to ensure that customers buy their products (thereby becoming "users"), or they may take place within social or governmental contexts in which advocates of an innovation seek to ensure that users (actual or potential) "buy" their ideas. Users can be both a market for products or ideas and a source of ideas and products in their own right. Users can also co-create products and ideas with firms or with other users. The involvement of users in innovation may be carefully managed, planned and ordered or it may be spontaneous and hard to control, with users creating their own rules of engagement. Certain forms of user innovation can lead to the most fundamental changes for organisations, markets and for public policy. This edited collection will provide insights into many aspects of the user innovation phenomenon.

Perspectives on users in the innovation literature

The Innovation Studies literature has evolved from an initially overwhelmingly supply-side perspective in which users possessed needs (e.g. Rothwell *et al.*, 1974), were "tough customers" (Gardiner and Rothwell, 1985), or "lead users", (von Hippel, 1986), all of whom may be harnessed to benefit firm innovation processes. This literature has developed to explore many non-traditional sources of innovation, for example communities (Franke and Shah, 2003), hackers (Flowers, 2008), open-source (Lakhani and von Hippel, 2003). It has also explored how firms can actively seek to prevent users from innovating (Braun and Herstatt, this volume). However, the literature has tended to retain its supply-side perspective.

It has also been argued that the process of innovation is becoming democraticized as improvements in ICT enable users to develop their own products and services (von Hippel, 2005). That users will often freely share their innovations with others, termed free revealing, has been widely documented (e.g. Allen, 1983) and this forms a key element in the rapid dissemination of certain forms of user-led innovation. The potential for users, either as individuals or as groups, to become involved in the design and production of products has clearly been recognised for some time. However, these conceptions of user-supplier relations in innovation all tend to depict a relationship in which the supplier is able, in some way or another, to harness the experience or ideas of users and apply them to their own product development efforts.

In contrast to the innovation studies literature, the Science and Technology Studies (STS) literature tends to adopt a more user-centric perspective, exploring how users actively shape technologies and are, in turn, shaped by them within the processes of innovation and diffusion. These processes are viewed as highly contested, with users, producers, policymakers and intermediary groups providing different meanings and uses to technologies (Oudshoorn and Pinch, 2003). The way in which design and other activities attempt to define and constrain the ways in which a product can be used have been viewed as an attempt to configure the user (Woolgar, 1991). Within this literature, users are seen as having an active role in seeking to shape or re-shape their relationship with technology, developing an agenda or "antiprogram" that conflicts with the designer, and going outside the scenario of use, or "script", that is embodied in the product (Akrich and Latour, 1992). Users lack of compliance with designers and promoters of products and systems, far from being viewed as a deviant activity, is positioned as central to our understanding the processes of innovation and diffusion.

Drawing on both of these bodies of literature, it is clear that users can play a series of important roles in the creation, development, implementation and diffusion of technologies. Arguably, the boundary between producers and consumers of technologies has become less distinct and users play important roles throughout the entire innovation process, potentially developing or extending technologies or applying them in entirely novel and unexpected ways. In this situation the boundary between consumer and producer, or between "users" and "doers" (Castells, 1996) becomes harder to discern. Innovation becomes far more open (Chesbrough, 2003), and democratized (von Hippel, 2005), as well as more complex. Users may be drawn into the traditional "linear" model of innovation, but some forms of user activity may represent the emergence of a parallel or alternative system of innovation that does not share the same goals, drivers and boundaries of mainstream activity. The processes of innovation, diffusion and re-innovation are becoming increasingly complex and contested. This has potentially significant implications for our understanding of innovation as a whole.

More pragmatic and largely empirical approaches to user involvement in innovation may be found in the innovation management literature. This body of literature deals with the problems faced by organisations seeking to involve users in some aspect of innovation like design, usability or diffusion. Although this body of literature draws heavily on both Innovation Studies and Science and Technology Studies, it more concerned with the tools, techniques and methods by which users can be beneficially drawn into the processes of innovation. Examples of this literature include methods to enable firms to identify lead users and draw on their ideas (Herstatt *et al.*, 1992), how firms may shift innovation to users via toolkits (Franke and Pillar, 2004), how the internet can be used to draw users into product innovation (Sawhney *et al.*, 2005) and the role of user communities in the commercialisation of products (Hienerth, 2006). This body of work is an important contribution towards translating the theoretical and empirical insights made within the Innovation Studies and Science and Technology Studies literatures.

Structure of the Book

This edited collection has been structured into three sections that examine and explore the roles users play in innovation, how users may be drawn into specific innovation processes, and new research issues and directions in what has become an expanding and fast-moving area. Each section contains three papers that have been selected to provide complementary perspectives on users' involvement in innovation, providing a series of theoretical, empirical, practical and policy insights.

Exploring the role(s) of users in innovation

This section presents different perspectives on the contingent factors that surround users involvement in innovation and begins with an exploration of the historical construction of user innovation within the Innovation Studies literature. Looking back over several decades of work in this area, and taking as her starting point the seminal contribution of von Hippel, Voss takes a fresh look at the development of this large body of work. User innovation has been observed in many contexts over many years and is generally recognised to be widespread and potentially of interest to policy makers. Voss traces how social, economic and political contexts have shaped the development of user innovation literature, the locations in which user innovation has been examined and the trajectories of subsequent research. She argues that user innovation is far from being a consistent phenomenon and can be viewed as a collection of practices from disparate industrial sectors and social spaces. The importance of historical context in user innovation is also explored. The difficulty of characterising an entire phenomenon from such a disparate body of literature is explored and Voss argues for a more subtle analysis of user innovation that considers its temporal nature and the specificities of the contexts within which it has been observed.

Raasch, Herstatt and Lock, coming from an Innovation Studies perspective, focus on the temporal aspects of user innovation in their examination of the evolution of innovation in the field of sports equipment. Using the empirical context of a high performance sailing boat, they analyse innovation activities over several decades and find that a large proportion of innovations can be traced to modifications undertaken by users. Raasch, Herstatt and Lock analyse free revealing behaviour in this domain and highlight the importance of "high bandwidth" oral communication in this process. They find that the level of user innovation activity does not follow a simple unidirectional trend, but develops as the opportunities for innovation change over time. In the context of high-performance sailing they find that standardised solutions produced by manufacturers are often unable to reproduce the performance improvements that are achieved by users, with the results that they become outdated. Their results suggest that, given a stimulating setting, user innovation activity can be sustained over very long periods of time. Raasch *et al.* identify five drivers of user innovation which can be of great value to practitioners trying to gauge the dynamics of current user activity or the potential success of a future involvement of users in the process.

Stewart and Hyysalo explore the role of intermediaries in the development and appropriation of new technologies. Drawing on the Social Learning in Technological Innovation (SLTI) framework, they focus on those intermediaries that facilitate user innovation and its links to supply side activities and develop a framework to explore how intermediaries work in making innovation happen. Their primary concern is to better understand how innovation intermediaries engage in configuring, facilitating and brokering technologies, uses and relationships in uncertain and emerging markets. The authors explore the range of positions and the influence that intermediaries are able to assume, the influence they are able to wield in different innovation contexts, and how they are able to bridge user-developer innovation domains. They also examine in depth how intermediaries affect the shape of new innovation and communication technologies and the importance of nurturing the user-side intermediaries that are crucial to an innovation's success. In their work Stewart and Hyysalo identify five cross-cutting issues: the presence of an ecology of intermediaries; the pivotal importance of uncertainty and learning; that the context of an innovation will produce important differences in the part played by intermediaries; and the importance of identifying and nurturing appropriate intermediaries.

Drawing users into the innovation processes

This section explores how users can be actively drawn into innovation processes by organisations and the roles of other actors in this process, beginning with an exploration of the part that may be played by a type of user who may point the way to future market demand — lead users. Bilgram, Brem and Voigt draw on the Innovation Studies literature to discuss the process of identifying and integrating "lead users" and explore the ways in which lead-user selection processes can be developed to take advantage of Web 2.0 applications. It has long been recognised that innovation is often an uncertain process and that innovations that are fundamentally new can have a high failure rate. In this chapter Bilgram, Brem and Voigt explore the potential for drawing lead users into new product development processes in order to mitigate the risks of such innovations.

Janssen and Dankbaar also examine the issue of consumer involvement in radical innovations. They present a detailed analysis of the requirements of consumer involvement in different situations, covering the two main phases of the development process ("discovery" and "incubation") and specified for three types of radical product innovation ("technologically really new", "trend-break really new" and "breakthrough"). Drawing on the Innovation Management literature they identify a mix of differentiating characteristics for six forms of new product development, presenting a model to select the appropriate techniques, before evaluating the model using an historical comparative case study approach. Janssen and Dankbaar identify 20 techniques for proactive user involvement and outline requirements for the involvement of users in six situations associated with different forms of innovation. Their work indicates that, in the context of product innovations, the information generated by the involvement of consumers is one of the most important factors in giving direction to radical product innovations. The authors argue that new product developers seeking to involve consumers should take great care in selecting the technique that will most appropriate for the outcomes they wish to achieve.

Moors, Boon, Nahuis and Vandeberg explore user-producer interactions in emerging pharmaceutical and food innovations and explore the issues of demand articulation and interactive learning. Drawing on both the Science and Technology Studies and the Innovation Studies literatures, they show how organised user-producer interactions via intermediary user organisations or consortia are important tools for articulating demand and facilitating learning amongst patient organisations, researchers, and private and public organisations. The chapter focuses on developing a classification scheme for user-producer interactions and argues that such interactions will vary according to the phase of technology development, its flexibility and the heterogeneity of the user population. The classification scheme developed by Moors, Boon, Nahuis and Vandeberg offers the potential to both evaluate and improve the organisation and management of user-producer interactions in innovation processes.

New directions in UI research and policy

This section presents a series of papers that introduce novel areas of enquiry that open up new and important avenues for future research. Schulz and Wagner focus on 'outlaw community innovations' and present results from a large scale survey of two online outlaw communities based around Microsoft's Xbox. In their study Schulz and Wagner find that most outlaw users modify or hack their consoles in order to gain access to functions that have not been provided by the manufacturer and are largely motivated to modify their Xbox by the prospect of accessing pirated software. They are also find that, in this context at least, outlaw users are also motivated by the sheer fun of hacking computer systems. Drawing on Innovation Studies approaches they argue that whilst such users can enhance products, they also engage in pirate behaviour and that manufacturers will increasingly need to weigh up the costs and benefits for their businesses of the existence of such outlaw communities.

In contrast, Flowers examines the implications for innovation policy of user innovation and outlines how the processes, participants and dynamics of innovation are changing and the dominant 'linear' model of innovation is being reappraised. The paper draws on the Innovation Studies literature to explore the wider definition of innovation that is beginning to emerge and examines how policy makers are beginning to engage with user innovation. Individual initiatives in Canada, The Netherlands and the UK aimed at developing firm-level and consumer metrics of user innovation activity are outlined and Flowers summarises the results of these surveys. The emergence of a strand of user innovation research that focuses solely on policy as a distinct activity is introduced and the chapter points the way to the emerging research agenda in this area.

Braun and Herstatt explore how firms can actively seek to exclude user involvement and place barriers in the path of user innovators. Building on Innovation Studies approaches they argue that although there is growing evidence that firms are becoming more open and are using outside resources in their innovative efforts, suppressing user innovation is a widespread corporate activity. Braun and Herstatt examine the scale and shape of this 'anti' user innovation phenomenon and outline the corporate incentives for this behaviour. Case examples are presented and the authors explore the conditions under which such behaviour is likely to be sustainable and profitable, arguing that changes in the business environment make exclusionary strategies unlikely to succeed in the long term.

Taken together, the papers included in this edited collection extend an important and ongoing debate concerning the nature and significance of user innovation. The debate in academia is necessarily multidisciplinary and collaborative endeavours are opening up spaces for novel methodological approaches and the development of new theoretical and conceptual tools and research agendas. This collection is intended as a contribution towards those endeavours. However, the collection is also aimed at managers and policy makers as well as academics and is intended to stimulate discussion both within and between these constituencies so that improved understandings of the processes that underpin and surround user innovation can contribute to improved practice and policy in this exciting and expanding field.

References

Akrich, M and B Latour (1992). A summary of convenient vocabulary for the semiotics of human and nonhuman assemblies. In: *Shaping Technology/Building Society: Studies in Sociotechnical Change*, WE Bijker and J Law (eds.), pp. 259–264. Cambridge, Mass.: MIT Press.

Allen, RC (1983). Collective invention. *Journal of Economic Behaviour and Organization*, 4, 1–24.

Castells, M (1996). The Rise of the Network Society. *The Information Age: Economy, Society and Culture*, Vol. 1. Mass: Blackwell.

Chesbrough, HW (2003). *The Era of Open Innovation*. MIT Sloan Management Review, Spring.

Franke, N and S Shah (2003). How communities support innovative activities: An exploration of assistance and sharing among end-users. *Research Policy*, 32(1), 157–178.

Gardiner, P and R Rothwell (1985). Tough customers: Good designs. *Design Studies*, 6(1), 7–17.

Franke, N and F Pillar (2004). Value creation by toolkits for user innovation and design: the case of the watch market. *Journal of Product Innovation Management*, 21, 401–415.

Herstatt, C and E von Hippel (1992). From experience: Developing new product concepts via the lead user method: A case study in a low Tech "field". *Journal of Product Innovation Management*, 9, 213–221.

Hiernerth, C (2006) The commercialization of user innovations: The development of the rodeo kayak industry, *R&D Management*, 36(3), 273–294.

Lakhani, KR and E von Hippel (2003). How open source software works: "Free" user-to-user assistance. *Research Policy*, 32(6), 923–943

Oudshoorn, N and T Pinch (2003). How users and Non-users matter. In: *How users Matter. The Co-Construction of Users and Technologies*, N Oudshoorn and T. Pinch (eds.), Massachusetts: MIT Press.

Rothwell, R, C Freeman, P Jervis, A Horsley, AB Roberston and J Townsend (1974). SAPPHO-Updated; Project SAPPHO Phase II. *Research Policy*, 3(3), 258–291.

Sawhney, M, G Verona and E Prandelli (2005). Collaborrating to create: The internet as a platform for customer engagement in product innovation. *Journal of Interactive Marketing* 19(4).

Von Hippel, E and R Katz (2002). Shifting innovation to users *via* toolkits. *Management Science*, 48(7), 821–833.

Von Hippel, E (2005). *Democratizing Innovation*. Cambridge, Mass.: The MIT Press.

Von Hippel, E (1986). Lead users: A source of novel product concepts, *Management Science*, 32(7), 791–805.

Woolgar, S (1991). Configuring the user: The case of usability trials. In: *A Sociology of Monsters: Essays on Power, Technology and Domination*, J Law (ed.), London: Routledge.

Part I

Exploring the Role(s) of Users in Innovation

THE HISTORICAL CONSTRUCTION OF USER INNOVATION

GEORGINA VOSS
University of Brighton, UK

Introduction

The user innovation (UI) literature has developed over three decades to describe phenomena such as lead users, sticky knowledge, free revealing, horizontal innovation communities, and toolkits. Following the advent of the internet user innovation has become more visible, moving beyond the academic and industrial arenas to attract the attention of policy makers (e.g. DCTI 2007, DTI 2004, DIUS 2008, NCM 2006, BIS 2009).

Beyond theoretical constructs, one frequent comment made around the UI phenomena is the *diversity* of places in which it has been observed, which include medical and scientific instruments, outdoor sporting goods, and digital products. The implications of this observation are twofold. Firstly, because UI has been observed in so many distinct domains, it is possible that it is more generally widespread and thus of interest to policy makers who wish to qualify its economic and social impacts. Secondly, because UI has been documented in these domains in the past, it will continue to occur there in the future and policy efforts should be targeted towards these sectors.

The aim of this chapter is to deconstruct these assumptions by mapping and describing the historical construction of the user innovation phenomenon by tracing its development in the innovation literature against the historical backdrop in which the work was being conducted. The purpose of this is to begin to examine how social, economic and political historical contexts have shaped the development of the field, the locations in which UI was explored and the subsequent trajectories of research.

The analysis is influenced by Godin's work on the historical construction of the linear model of innovation (Godin, 2006). By tracing the history of the model to the present day, Godin demonstrated that rather than arising from a unified body of work, the linear model was incrementally constructed by three different bodies of actors.

Each group developed aspects of the model which were relevant to their own interests; thus its development is intrinsically tied to the context of these actors' activities.

In examining the development of a particular subfield of innovation studies this chapter also aims to respond to the critique that although the field is nearly four decades old, "little is written on innovation studies as such" (Fagerberg and Verspagen, 2009), not least given that the history of a technology is contextual to the history of the industrial studies associated with it (Dosi, 1982). The historical context in which the issues were studied therefore influenced the trajectory of the development of the field, and this is particularly true for the development of user innovation.

Research on aspects of user innovation is not confined to innovation studies (Flowers and Henwood, 2008). Oudshoorn and Pinch (2003, 2008) describe other areas which have examined the relationship between users and technology.[1] However, this chapter focuses on the body of work which has arisen from Eric von Hippel's work in the 1970s (von Hippel, 1976, 2005) which challenged the dominant view in innovation studies at the time that product innovation was conducted by manufacturing firms. There are two reasons for focusing on this section of the work, which has developed primarily within the innovation studies literature. Firstly this work has specifically focused on the user-as-innovator (rather than the broader relationships between users and technologies) and has arisen from and developed within a specific arm of the innovation canon. Secondly, this particular body of research has been central in informing the recent science and technology policy debates about user innovation, and the broader definition of innovation (Flowers, this volume). Limitations or assumptions made in this literature may therefore be carried over into subsequent policy activity.

Following Godin (2006) this chapter maps out the development of the user innovation literature in four themes across different sectors. In the first theme, work in the early 1970s focused on scientific instruments and machinery at a time when these fields were being widely examined in the nascent innovation studies field, due to their importance to the UK and US economy. The second theme focused on medical instruments developed by individual practitioners. Here I suggest that whilst the practices of medicine may be largely unchanged over this time in terms of

---

[1]These include the Social Construction of Technology (SCOT) approach which focuses on the role of users as agents of technological change (Kline and Pinch, 1996; Bijker and Pinch, 1987); feminist studies of technology which examines the role of women in the development of technologies, and power relations between diverse actors (e.g. Cockburn and Ormod, 1993; Oudshoorn et al., 2004); semiotic approaches which focus on "scripts" and "configuration of users" (Akrich, 1992; Akrich and Latour, 1992; Woolgar, 1991); and media and cultural studies, in which the role of users in making, distributing and consuming cultural products has been a central aspect of the field since its inception (Jenkins, 2006a, b; Silverstone et al., 1992).

skills and interactions, the regulatory environment has altered meaning that the user activities described in earlier work may not be so achievable now. In the third theme beginning in the late 1990s, a research cluster focused on the role that users played in developing modifications around outdoor sporting equipment such as rodeo kayaks and skis. Much of this work has been conducted by researchers in regions where participation in such sports is high, and this body of work also seems to correspond to wider trend of the social democratisation of sport. The most recent and prodigious theme in research corresponds to the advent of the internet and the wider "democratisation of innovation" (von Hippel, 2005), and focuses on the development of digital products and more latterly, digital tools. However, unlike Godin's analysis which occurs temporally, I do not propose a linear model of user innovation research. The themes described do not correspond to consecutive periods which feed directly into each other, but overlaying layers which frequently occur simultaneously (i.e. research around sporting goods and digital products and tools began at roughly the same time) and can be extended and elaborated on by multiple actors.

The chapter also does not focus overtly on the development of key theoretical concepts or the tight-knit social network of researchers in the UI community. Instead, it examines the historical context of the phases, the justifications offered for examining specific sectors in which UI activities occurred, and offers challenges as to why it is not sufficient to simply say "User innovation occurs in these sectors".

Theme 1: Scientific Instruments and Machine Tools

The first theme of user innovation research was heavily influenced by the concurrent development of the nascent field of innovation studies which arose in the UK and US out of the post-World War II government drives to boost economic growth (Fagerberg and Verspagen, 2009). In the US between the 1950s and 1960s, government policy was heavily influenced by concepts of "science push" (von Tunzelmann et al., 2008), drawing theoretical justification from the work of Nelson (1959) and Arrow (1962). The post-war industries which had developed included computers, semi-conductors, and biotechnology, with new firms in this area acting as important actors with respect to R&D activities in the US (Bruland and Mowery, 2005).[2] Following the oil shock of the 1970s and associated concerns about the adequacy of energy supplies for economic growth (Rosenberg, 1982, p. 81), UK policy emerged to follow market pull theory, following Schmookler's (1962, 1966) work on the role of demand in providing economic incentives for innovation. This

---

[2] These sectors were in turn influenced by a weak patent system and liberal licensing, high levels of basic research in universities, and advances in electrical technology (as affected by government procurement). Biotechnology was also supported by federal expenses in R&D.

period also coincided with the changing post-war economies of both the US and the UK, with the latter increasingly focused on the chemicals and electrical industries, and automobiles (Rosenberg, 1982, p. 256).

The drive in innovation research arose from the need to understand innovation for government intervention, particularly in the UK as recession of the 1970s developed (Mowery and Rosenberg, 1979). With the economic downturn, there was growing concern with the problems of slower growth in productivity and income (amongst other factors), so new empirical studies of innovation were funded by government agencies; and earlier work was also "exhumed and re-interpreted" (Mowery and Rosenberg, 1979).

From its inception, innovation studies was intellectually cross-disciplinary (Fagerberg and Verspagen, 2009), focusing on contemporary technological and industrial issues around innovation processes in the first instance, rather than applying and testing established theoretical frameworks.[3] Accordingly, with an absence of theoretical frameworks to test, early research in the area focused on spaces in which innovative activity was known to happen, such as mechanical engineering and instruments, a practice reflected in von Hippel's seminal paper in 1976. This was the first work which explicitly articulated the role of users as active *innovators* (von Hippel, 1976) rather than as a source of input into the innovation activities conducted by manufacturers (Rothwell, 1994). By publishing in the newly-founded *Research Policy* this work aligned with the emergent innovation studies field.

Empirical evidence for the original 1976 paper was drawn from the field of scientific instruments, which were chosen as previous research had show that innovation in scientific instruments was predominant in responses to user needs (Utterback, 1971; Shimsoni, 1966).[4] Project SAPPHO (Rothwell, 1974), had also focused on innovations in the chemicals processing and the scientific instruments industries, differentiating between the process innovations of the former and the product innovations of the latter. von Hippel cited Project SAPPHO (Rothwell *et al.*, 1974) as being influential in demonstrating that characteristic patterns in the innovation process varied as a function of the type of good examined; and emphasising the importance of recognising user needs in the innovation process: "User needs must

---

[3] SPRU, the first research centre established to exclusively researching issues of innovation with respect to science and technology was deliberately cross-disciplinary, employing frameworks and materials from economics, history, political science and philosophy (Fagerberg and Verspagen, 2009).

[4] Utterback (1971) focused on Massachusetts firms in this sector as they had developed products comparatively recently (between 1963 and 1968), had received awards from "Industrial Research" for technical excellence, and indicated that the source of need for their innovations were from potential or actual customers outside the firm.

be precisely determined and met and it is important that these needs are monitored throughout the course of the innovation process as they rarely remain static" (Rothwell et al., 1974).

The industries which were studied in the 1976 paper were all recent, having had first commercialisation of instruments between 1939 and 1954; and the instruments studied were electron microscopy and nuclear magnetic resonance (NMR) machines. Whilst the manufacturing firms were based in the US (Varian Inc, Paolo Alto; Bloch, Stanford) and Germany (Siemens), the users were individuals based in North America.

Following the focus on scientific instruments in order to solve problems in use, von Hippel and Finkelstein (1978) examined user customisation around clinical chemical analysers, again influenced by the high levels of product innovation in the previous two decades. Details of the von Hippel's 1976 paper were re-examined in his 1978 paper on customer-active paradigms (von Hippel, 1978), in which he also drew on further research on the importance of customer requests in idea generation in chemical products (Meadows 1969), plant processes, equipment and technologies (Peplow, 1960) and standard industrial products (Robinson, 1967). The focus on industrial goods had also been weighted by previous work describing how industrial goods were produced in response to needs and specifications of engineers both in the US (Brand, 1972; Robinson et al., 1967) and the UK (Buckner, 1967), who make buying decisions. The importance of engineers in the 1960s industrial landscape was further emphasised by work describing their role in developing the specifications in the engineering design process (Allen, 1966; Marples, 1961).

von Hippel (1982) recognised that user activity would vary in intensity and importance across industries, drawing on secondary data from the 1960s in the UK (Taylor and Silberston, 1973). Data was again drawn from industries that were of specific relevance to policy-makers at the time, namely chemical, mechanical and electrical engineering.

Thematic aspects and unit of analysis

In this early stage of the UI work the primary objective was to describe the phenomenon and the manner it which it presented challenges to the manufacturer-centric paradigm, and to find examples of it in previous research. Emergent themes that arose from this period include the Customer-Active Paradigm (von Hippel, 1978), which would later be developed into lead user theory; and how appropriability of returns affects the locus of innovation (von Hippel, 1982). The unit of analysis was primarily the innovation itself, and the organisation in which the innovation occurred. Users here were individuals who were not employed by the manufacturing firms (Riggs and von Hippel, 1994).

Other issues

After the initial focus in the 1970s and early 1980s, little further empirical UI work around contemporary scientific instruments and machine tools was conducted around the sector. Instead, conclusions about the UI phenomenon in that sector were drawn using data from the late 1960s to the early 1980s. This included Riggs and von Hippel's work on UI around spectroscopic instruments (1994) which focused on major improvements on instruments produced commercially by *Auger* and *Esea* before 1988, the majority of which were developed between 1969 and 1982. Justification for the dataset and sector analysis was previous work demonstrating innovation in the area, and Rigg's background in the area. Similarly, in developing the initial theoretical concepts around lead users, von Hippel (1986) cited empirical data from the 1960s and 1970s (Freeman, 1968; Berger, 1975). The focus on the role of users in innovation around capital goods was further emphasised by Rosenberg (1982, p. 122) who described how learning-by-using was more likely to be prevalent around these products as the performance characteristics of durable capital goods cannot be understood until after prolonged experience with them. This in turn was likely to be related to the relationship between learning-by-using and system complexity, embodying high fixed costs and long lead times in which design characteristics emerge (Rosenberg, 1982, p. 135). Beyond this, little further work described *why* UI should occur around machine tools and scientific instruments. Accordingly, justification that UI happened around these sectors was based on evidence collected at a time when the sector was economically strong and thus of interest to innovation studies academics and policy makers.

Theme 2: Medical Instruments

First work in the area

The second theme of the user innovation research focuses on innovation around medical instruments. The involvement of users in healthcare services and technologies has been extensively examined by the STS community (e.g. Van Kammen, 2005; Ourdshoorn *et al.*, 2005; Woolgar, 1991). In the UI field however, this focus was driven by von Hippel's original paper (von Hippel, 1976). Of the 111 first-to-market innovations examined in the 1976 article, several were medical instruments; and von Hippel described in the conclusion how further "anecdotal evidence" indicated that user innovation was likely to occur in this sector (von Hippel, 1976).[5]

---

[5] Whilst the clinical chemical analysers examined by von Hippel and Finkelstein (1978) were used by individuals in hospital settings, they were predominantly utilised by technicians in clinical chemistry laboratories rather than physicians and as such could be considered as scientific rather than medical

This provided justification for Shaw (1985) and Biemens (1991) to examine interactions between users, manufacturers and third-parties around innovation of medical equipment; which in turn provided justification for Lettl's work on users and radical innovation (Lettl, 2007; Lettl *et al.*, 2006a, 2006b; Lettl and Gemunden, 2005), work on professional users (Chatterji *et al.*, 2008), and research on how manufacturers can limit user innovation activities (Braun and Herstatt, 2008).

Thematic aspects and unit of analysis

As such, medical instruments became a known site of innovation for researchers focusing on the area, allowing development of case studies of medical equipment as sites of theoretical relevance. These included contributions to the early stage of new product development (Lettl *et al.*, 2006; Lettl, 2007) and user entrepreneurship (Lettl and Gemunden, 2005). The units of analysis were the user innovations, and individual medical practitioners who innovated in order to solve problems associated with moving beyond the limits of conventional technology that constrained their day-to-day work (Lettl, 2007).

Justification

In justifying their focus on innovation by physicians Chatterji *et al.* (2008) also cited historical instances of user innovation around medical instruments, including the Fogarty catheter (Gassman, 2006), the Pruitt-Inahara Carotid shunt (Gales, 2006; cited in Chatterji *et al.*, 2008), coronary shunts and spinal cages. Beyond historical material and contemporary anecdotal evidence that "the level of innovative activity was high in this industry" (Shaw, 1985) researchers have also attempted to explain *why* user innovation may be likely to occur around medical instruments. Shaw noted that before being introduced into clinical use, medical equipment would first need to undergo clinical assessment and trial, where " 'state of the art' clinical and diagnostic knowledge resides with in the user" (Shaw, 1985); and the prime characteristic of the medical equipment innovation process was the multiple and continuous interactions between manufacturers and users. Noting previous work describing the phenomena, Chatterji *et al.* (2008) described how the physical and ongoing nature of physicians' work means that when they use medical devices during procedures they will "often imagine ways in which the product could be made better"; and also hypothesised that user innovations should be more prevalent in market environments with strong IP rights and non-specialised complementary assets (such as the market for medical instruments in the US).

---

instruments. The focus on user innovation around medical instruments was therefore implied to be some quality of the user context and experience of medical practitioners.

Unlike the later focus on extreme sporting goods and online products, few of the studies on UI in medical devices justified their focus through market size; with the exception of Chatterji *et al.* (2008) who emphasised the size of the medical device industry in the US as being of importance, comprising 6000 companies.

Issues of generalisability

Whilst these aspects of use of medical instruments may remain constant as an inherent part of physicians' work, other factors are more context specific which makes it difficult to generalise from this smaller body of research that user innovation around medical instruments is a given phenomenon. Firstly, with the exception of Chatterji *et al.* (2008), the work focused on medical professionals who work in public teaching and research hospitals, and universities attached to hospitals. These institutions were based in the UK (Shaw, 1985), the Netherlands (Biemens, 1991), Germany (Braun and Herstatt, 2008; Lettl, 2007; Lettl *et al.*, 2006a, 2006b; Lettl and Gemunden, 2005) and Switzerland and France (Lettl, 2007; Lettl *et al.*, 2006a, 2006b; Lettl and Gemunden, 2005). By contrast Chatterji *et al.* (2008) examined US-based physicians, the majority of whom worked in group or private practices (31% and 24% respectively), with only 8% working in medical schools. Given that much of this work emphasises the importance of the network in which physicians are embedded for any innovation, these different contexts can differentially influence the dynamics of such innovation. For example, the role of national and international government bodies such as the UK Medical Research Council (MRC), the UK Department of Health and Social Security (DHSS), the UK Department of Industry (DoI) (Shaw, 1985) and the European Union (Lettl *et al.*, 2006a, 2006b) is described, but there is clear variation in institutions and infrastructures between countries.

Secondly, the historical context of this sector with regard to changing technologies and regulations have not been examined. The empirical data on medical instruments is drawn from a time period spanning the late 1960s to the late 1980s (Shaw, 1985; Biemens, 1991), patents registered in the US between 1990 and 1996 (Chatterji *et al.*, 2008); and case-studies between the mid-1980s and the late 1990s (Lettl *et al.*, 2006a, 2006b; Lettl, 2007). The changing nature of medical devices in this time has been noted with anecdotal data provided about "a number of radical innovations [which] have emerged just recently with new communication and information technologies finding their way into the operating table" (Lettl and Gemunden, 2005). One key aspect of this context is the standards and regulations to which medical devices must conform. These standards and regulations, which are aligned between the EU and the US, have changed considerably from the mid-1990s onwards with the result that it has become more difficult to design devices which can be validated (Alexander and Clarkson, 2000). These additional regulatory barriers may hinder

UI activities, or make them the province of fewer medical practitioners who have the necessary resources to innovate around these barriers.

The UI research around medical instruments has remained thus far the province of a small body of researchers.[6] Unlike the work on scientific instruments and machine tools, the focus on medical instruments was not precipitated by governmental policy focus or intervention into the sector as a site of economic growth and importance; but instead due to the presence of the phenomena.

Theme 3: Outdoor Sports Consumer Products

Initial work in the area

The focus on industrial goods in the early UI research was acknowledged by von Hippel and Urban (1988) and the third phase on UI in outdoor sports consumer products represented a shift towards consumer goods. Informal evidence was given about users also innovating around consumer goods by Urban and von Hippel (1988) who briefly described users who added eggs to their shampoo; the development and modification of all-terrain bikes by users in Northern California prior to their wider commercialisation; and the role of athletes in developing sporting products including skis, running shoes and tennis rackets.

This phase of empirical work on UI in consumer goods was initially lead by Sonali Shah, who focused on UI in outdoor sporting equipment in skateboarding, snowboarding and windsurfing, and open source software for her doctoral thesis (Shah, 2003). As with work on medical and scientific instruments, these sports were chosen as known sites of UI, whereby innovations within the sporting community began amongst the enthusiastic users before migrating to the mass market (Shah, 2000). Shah also described how UI was also likely to happen around sporting goods as the significant size of the market meant that users should have an interest in participating in the development of the sports "because of the attractiveness of the activity". Whilst mountain biking and rollerblading also fulfilled similar characteristics, the three selected sports were chosen as the student researchers were familiar with them.

As the first empirical work to focus on UI in consumer goods, Shah's work was highly influential. In particular, the working paper produced at an interim stage of her studies which focused exclusively on innovations in sporting equipment (Shah, 2000) has been frequently cited by the further researchers who also focused on outdoor sporting goods (e.g. Luthje *et al.*, 2005; Hierneth, 2006; Baldwin *et al.*, 2006).

---

[6]Whilst Lettl has published most widely and recently around it, the same dataset of five cases-studies of medical technologies has informed all of these publications (Lettl, 2006; Lettl *et al.*, 2006a, 2006b; Lettl and Gemunden, 2005).

Thematic aspects

The focus on sporting goods allowed development of UI theory around a number of key concepts. Unlike the established markets of scientific and medical instruments, UI around sporting activities created new markets and equipments for sporting goods. These markets had also developed comparatively recently — mountain biking activities emerged in the 1970s (Luthje et al., 2005), snowboarding in 1965 (Shah, 2000), windsurfing in 1964 (Shah, 2000), kitesurfing and rodeo kayaking in the 1970s (Hierneth, 2006); and were sizeable commercial industries by the mid to late-1990s. This allowed researchers to collect material from users who had been involved in the emergence of the products from their inception and throughout their subsequent development. Accordingly, work in the area examined commercialisation of consumer goods from initial UIs (Baldwin et al., 2006); and community reactions to the transition from small scale product development to users supplying products for a new market system (Hierneth, 2006).

The type of sporting activities examined in this phase were also those in which participants were frequently part of a community of practitioners (rather than solo actors as in prior UI activities) — Hierneth (2006) describes how rodeo kayakers form communities as participants paddle together for safety, discuss their activities, and form teams to compete in the Olympics. Unlike previous UI work on scientific and medical instruments, users of sporting goods were also non-professionals (apart from professional sportspeople) who were unlikely to have the same access to financial and organisational resources as their professional counterparts.

Researchers accordingly began to focus on the supportive role of the community for user innovators, in terms of free revealing (Franke and Shah, 2003). The "young and trendy" nature of these newer sports such as kite-surfing meant younger users were prevalent, and formed communities in both online and offline spaces thereby allowing research into how user communities of support and discussion operated online (Franke et al., 2006). Hierneth (2006) suggested that the reasons for this development in this particular space were because of the low costs in developing and testing new products, and that the designs in rodeo kayaking could be changed by (non-professional) users with the necessary skills.

Concepts which had already been developed in the earlier UI literature were also examined, including the role of lead users for product adoption and diffusion (Schreier and Prugl, 2007; Shah, 2000).

Justification

The size of these new sectors was also emphasised to demonstrate that markets which emerged from UI activities could be financially significant, providing further validation for the work. In 2000 total retail sales for bikes in the US were $5.87 billion, of which 65% were from mountain bikes (Luthje *et al.*, 2005). 'Paddling activities' (which encompassed rodeo kayaking) was the second fastest growing outdoor sport in the US (Hierneth, 2006); and by 2002 rodeo kayaking was a $100 m industry in the US (Baldwin *et al.*, 2006). Whilst these activities began in the 1970s, historical accounts of the growth of the sector were not produced until the late 1990s (Penning, 1998) and research into the sector commenced shortly afterwards (Franke and Shah, 2003; Luthje *et al.*, 2005). Similarly, rodeo kayaking had developed from a hobbyist pastime dominated by UIs in the 1970s to a commercial industry in its own right by the mid-1990s (Hierneth, 2006), increasing in size from 7,000 US enthusiasts in the 1970s to 435,000 in 2002 (Baldwin *et al.*, 2006).

Beyond the nature and size of the sporting communities themselves, geography and locality also played an important role in determining the types of sports which UI researchers investigated in this period — namely, those which occurred around local or national region of the researcher's universities. Shah's original work (Shah, 2000) was conducted at MIT in Cambridge, MA, and focused on sports which were developed in North America. Her later collaboration with Nik Franke from Vienna University of Economics and Business Administration (Franke and Shah, 2003) brought in further data on extreme sports which were specific to the local area including canyoning in the Alps. Several of the researchers who subsequently focused on UI in sporting communities were also based in universities in Germany and Austria, and much of the empirical data was collected from Northern European sporting communities (e.g. Luthje, 2003; Hierneth, 2006; Franke *et al.*, 2006; Schreier and Prugl, 2007).[7]

Unit of analysis

Accordingly the unit of analysis for the much of the work within this phase was on non-professional individuals, and the communities which support them. A recent exception is the work on the role of intermediaries in mediating UI activities, examining the role of sporting goods retailers (Luthje and Franke, 2003), which were however chosen because "salespeople [in sports shops] are often very active in the sports themselves".

---

[7] Christian Luthje from the University of Hamburg-Harburg, Germany, focused on individuals who participated in the four sports in Germany which were most cited in the outdoor sporting journals: climbing and mountaineering, hiking, cross-country skiing and mountain biking (Luthje 2003). Christoph Hiernerth from Vienna University of Economics and Business Administration, examined international rodeo kayakers entering competitions in Graz, Austria (Hierneth 2006). Both Franke and Martin Schreier (also from the Vienna University of Economics and Business Administration) examined kite-surfing in online communities based predominantly in Germany, Greece, Austria and Switzerland (Franke *et al.*, 2006, Schreier and Prugl 2007).

This was a discrete phase of the research which coincided with the development of work on online products and platforms — as mentioned, Shah's doctoral thesis was on both sporting goods and open source software. Accordingly, the focus of the work shifted from the physical communities and their activities to online spaces where users are able to use toolkits and other digital tools for UI activities (Fuller *et al.*, 2007).

Phase 4: ICTs and Digital Products and Tools

The fourth phase of the UI research focused on digital products, platforms and tools. This represented a key shift in the literature — from focusing on instances of UI in particular industrial and hobbyist spaces, research was now able to examine UI across a multiplicity of sectors. This shift in focus was influenced by the recent advent of the internet in the late 1990s and the associated rise of open source software development, rather than the earlier role that users had played in the development of the nascent computer industry.

As with outdoor sporting equipment, users played a key role in the development of both software and the industry which developed around it. Rosenburg had acknowledged the importance of users in the development of the computer industry in the 1960s and 1970s, desribing how the development of effective software was highly dependent on user experience (Rosenberg, 1982, p. 139). Knight's unpublished PhD thesis described the role of users in the evolution of digital computers between 1964 and 1962 (Knight, 1963, cited in Urban and von Hippel, 1988), material from which was frequently cited by UI researchers to show the functional locus and frequency of user innovations. In the UK, the involvement of users and suppliers in software innovation was classified by Voss (1985); who also noted the do-it-yourself learning that was inherent in this process, and the "familiarity with computer technology rather than technical learning" which allowed user engagement. The role of users in the development of the first business computer (Caminer *et al.*, 1996) and in the development of the US software and hardware industry via "hacker culture" (Turner, 2006) was also described from a historical perspective.

Until the advent of the internet, however, research on UI around digital products and platforms was not embedded in this historical context but instead focused on individual instances of UI activity (in a similar manner to the work done on medical and scientific instruments). Cases were chosen as they occurred in growing industrial spaces with substantial economic impacts. Urban and von Hippel (1988) examined user contributions to product marketing for Computer-Aided Design (CAD) systems, selecting the case because it described a "large, growing and rapidly changing market".[8] Later work on toolkits (von Hippel, 2001b) described a more active role

---

[8]The sector comprised of 40 firms in a $1bn market, growing at 35% between 1982 and 1986.

of users in the innovation system for application-specific integrated circuits (ASIC) and computer-assisted telephony integration systems (CTI), again selected for study due to their size and rapid growth in the early to mid 1990s.[9] Morrison *et al.* (2000) examined the characteristics of individual Australian lead users around the computerised Online Public Access (OPAC) system. Whilst noting that "OPACs were initially developed by advanced and technically sophisticated users" in the US in the 1970s (with the US Library of Congress government funding), it was not made explicitly clear why OPACs have been selected as a site in which to examine the characteristics of innovating users. In the mid to late 1990s the increasing prevalence of digital products in society and industry was noted, with Ogawa (1998) acknowledging that digital systems had been introduced into Japanese paper-based inventory management systems as a result of his research into the locus of sticky knowledge in UI.

The emergence of internet technologies and the OSS movement marked a resurgence in UI research. Following its inception in 1985, the OSS movement broke into the public and academic consciousness in the late 1990s following a series of profile-raising activities, including the renaming of the movement from "free software" to "open source" in 1998 (Perens, 1988, cited in von Hippel and von Krogh, 2003). Two non-academic books were published shortly afterwards — the monograph *The Cathedral and the Bazaar* by Eric S Raymond (1999); and *Open Sources: Voices From The Revolution*, a collection of essays edited by Chris DiBona, Sam Ockman and Mark Stone (1998) — which described the movement from the perspective of its participants.

The potential for UI research from OSS activities was highlighted shortly after the movement's emergence into the mainstream by von Hippel, who described how although there were surface similarities between the two activities, "UI communities existed long before the advent of open-source software and far beyond it" (von Hippel, 2001a), thereby characterising OSS as a subset of UI. The similarities between OSS and UI were also made explicitly clear by von Krogh and von Hippel (2006, p. 976) who described how:

> "... in open source software projects we see that users are indeed major innovation contributors. Spurred by this we can rethink underlying examples about innovation. We then discover countervailing advantages available to user-innovators that can offset the manufacturers advantage of potential market scale: better information on emerging market needs, and a certain, even though sometimes small, internal market for what they develop."

---

[9] ASIC had sales of $13.5bn in 1994, accounting for 15% of all IC sales.

Research around UI and OSS also emerged in similar manners — examination of an unusual empirical phenomenon that challenged the manufacturer-centric paradigm of existing innovation frameworks. However, whilst OSS and UI bore similarities in the patters and motivations of user activities, the emergent bodies of policy activity, industry response and literature diverged. This was primarily due to the scale of OSS — by 1998 it was a movement comprising hundreds of "members" forming a community with its own "leading voices" and political ideology (DiBona et al., 2006, p. xxvi). Industry was already trying to engage with the phenomenon — Microsoft's 1997 'Halloween Memo' showed the company to perceive OSS as a competitive threat, and by 1998 IBM had already provided initial backing for Apache. Accordingly, the literature on OSS described a wide-scale phenomenon concerned with political and economic issues around proprietary software and systems of ownership and distribution; and research into the area focused on areas of motivations, governance and dynamics (von Krogh and von Hippel, 2006).

This contrasted to the UI research which had primarily focused on single case-studies in specific sectors. As a legacy of the ground-up manner in which the OSS community had formed and publicised its existence, much of the writing which described the phenomenon was conducted by OSS participants who described the movement and its activities in their own terms (e.g. DiBona et al., 1999). This contrasted with the academic legacy of UI which had emerged from the greater literature of innovation studies, and was led by economically framed discussions about the dynamics of technological innovation and the manner in which innovations could be appropriated for firms' benefits (rather than a user perspective).

Thematic aspects

Some researchers used UI theoretical frameworks to examine aspects of OSS, including around free revealing (Harhoff et al., 2003; Henkel, 2006); and toolkits, "unsticking" of information and heterogeneity of user needs (Franke and von Hippel, 2003). Other research decribed how OSS resembled UI communities that had previously been examined: Lakhani and von Hippel (2003) explained how "innovation users are frequently innovators" in their research on OSS; and Dahlander and Magnusson (2005) described the other non-firm innovation communities that developed innovations and voluntarily diffused them.

Beyond these studies, OSS acted more as a totem for the UI research communities, demonstrating that UI activities could operate on a grand scale with widespread economic, social and political impacts. Accordingly, whilst OSS provided a focus for much of the earlier work on innovative users in online communities, the later research moved away from OSS to examine the theoretical constructs developed in the previous UI literature, and ask how other community activities could be utilised in development for both digital (Fuller et al., 2006; Gassman et al., 2006) and physical products (Fuller et al., 2007). Focus on digital products and platforms also allowed wider examination into the phenomenon of toolkits. This had been developed from work on mass customisation (Franke and Piller, 2004), where the digital tools developed by firms for customer use differed from the complex business-to-business tools which required precise technical knowledge for operation.

Industries

von Hippel described the increasing trend of the cross-sector "democratisation" of innovation which was driven by "steadily better and cheaper computing and communications" (von Hippel, 2005, p. 177). Yet several studies of digital UI focused on sectors which had previously been examined in the UI literature — namely sports and outdoor equipment — as internet networking technologies greatly increased the scope of UI activities by enabling users to share their enthusiasms, feedback and products via simplified methods of interactions (Fuller et al., 2006). Using similar rationale as previous studies of sporting goods, UI around basketball trainers in online communities was examined because basketball was a team sport, so users were likely to share their experiences; unlike newer digital industries, the shoe market was already established and competitive; and one of the authors had personal knowledge of basketball (Fuller et al., 2007).

Entertainment products and the creative industries after 2000. Prugl and Schreier (2006) described how early computer games had been developed by users in the US and UK. As with other digital products, development of the fledgling computer games industry occurred in the 1960s and 1970s, but was not mapped by industry or cultural historians until 2001 (e.g. Readman and Grantham, 2006; Schilling, 2003; Kent, 2001). Jeppesen focused on Westwood Studios in Germany in examining how firms relied on external consumer communities (Jeppesen and Molin, 2003) and then on how these firms could supply toolkits to complement other customer support activities (Jeppesen, 2005). Jeppesen justified this continued focus by noting that Westwood offered its first toolkit to users in 2001 and had two titles in the top 20 best-selling computer games in the US; and by 2001 had 50,000 players playing 500,000 games per week. Jeppesen and Fredericksen focused on online music communities, examining why users in online firm-hosted communities reveal their innovations to the firm (Jeppesen and Fredericksen, 2006). The case firm, Propellerhead, was chosen as it had a history of engaging customers into its product testing and development, following an outside group hacking their ReBirth product.

More broadly, UI research continued to examine highly diverse locations of UI activity around consumer goods which was enabled or accelerated by online platforms and tools. Ujjual (2009) argues that the reason why the "vast majority" of current studies about UI are drawn from online user communities is because the sources of data in those locations is prevalent, easily identifiable and easy to collect. The increasingly vast array of online users communities provides a range of UI locations to examine; and the comparative ease by which survey instruments can be applied in these settings also further enables a shift towards more quantitative methodologies, beyond previous small-scale surveys.

Conclusion

In this chapter I have argued that user innovation as examined to date consists of a collection of practices from disparate industrial sectors and social spaces which does not yet fully describe a coherent narrative. Although these practices are probably representative of a consistent phenomenon, they also are part of a broader narrative about innovation and use practices which is not examined in the current UI literatures. Each of these practices examined to date may represent an innovation type specific to that area with the differences characterised by facets such as the unit of analysis. Further, the activities which have been described in each stage are embedded in the specific historical, industrial and regulatory context in which they were examined.

The first phase of UI research focused on scientific instruments, drawing on primarily qualitative case-study analysis with the innovation as the unit of analysis (e.g. von Hippel, 1976; von Hippel and Finkelstein, 1978). This first phase has been concerned with describing the initial phenomena, exploration of theories around the Customer Active Paradigm, appropriabilty of innovations, and early exploration of the concept of lead users. Case-study methodologies were predominantly used (e.g. von Hippel, 1976). The second phase of UI research focused on medical instruments, again drawing on primarily qualitative case-study analysis with the innovation as the unit of analysis (e.g. Lettl et al., 2006); although small-scale surveys of physicians (Luthje, 2003) and patent datasets have also been recently used for quantitative analyses (Chatterji et al., 2008). Theoretical constructs examined in phase two include the role of the user in New Product Development, and user entrepreneurship. In the third phase, research around outdoor and sporting equipment complemented qualitative case-study analysis (e.g. Shah, 2000; Hierneth, 2006) with further small-scale surveys of individual users (Franke and Shah, 2003; Franke et al., 2006; Schrier and Prugl, 2008). The unit of analysis has been both the individual user and user communities, in addition to the innovation itself; accordingly, research shifted focus from the individual user to the behaviour of community, with issues around free revealing, community dynamics and the impact of commercialisation activities on the community examined.

Finally, following the advent of the internet and the rise of the open-source software movement, phase four has focused on ICTs and digital products. Constructs examine include the design and use of toolkits (Prugl and Schrier, 2006; Jeppesen, 2005), motivations for free-revealing and participating in user communities and community interactions with firms (Dahlander and Magnusson, 2003). The arrival of online networking technologies has also facilitated the more systemic, quantitative-based final phase of the field, as researchers have made use of online survey tools to collect data from large user communities, such as those developing Apache (Franke and von Hippel, 2003) and Linux (Hertel *et al.*, 2003; Henkel, 2006) software.

Despite drawing on large bodies of historical data in the earlier work the historical perspective for the data has not been fully acknowledged. As von Hippel and Urban (1988) note, the dynamics of UI are shaped by the real world experiences of the users and how these are manifested through their needs. This however implies that when the world experience changes through shifts in context, needs will change too and the innovation dynamics will alter. The location of user activities may have shifted from the locations described in the original work, particularly where industry structures and markets have evolved and standards and other regulations have been introduced (for example, see Raasch, Herstatt and Lock in this volume). In the case of medical instruments, regulatory barriers in both the EU and US have the potential to limit user innovation around these devices (Alexander and Clarkson, 2000). Non-firm user innovation around digital products does not always conform to regulatory standards and such innovations can often be clamped down on and deemed "outlaw", until it is either adopted by legitimate organisations or the legal standards shift (Flowers, 2008).

The disparate focus on UI results from the emergent nature of the field, and of innovation studies itself. As Ujjual (2009) observes, "the selection of the empirical setting for investigating the user innovation phenomenon are seen to be strongly influenced by the feasibility and ease of information and data collection, and the ease and practicability in identifying reliable and manageable sources." This in itself is no bad thing – it could be argued that in the development of an academic field of study it is better to begin by examining areas in which the phenomenon is known to occur to gain rich and detailed insight, rather than spinning out into lesser-known spaces. UI research initially prioritised rich description of UI and developed theoretical constructs with which to characterise the phenomenon. However, with the maturation of the field and the emerging interest of policy-makers, it now becomes more important to undertake systematic reviews to provide the evidence base necessary for policy development to complement and further the earlier work. Understanding the historical context of the first wave of UI research is essential in framing the next stage; for as social and institutional frameworks facilitate and accelerate certain waves of technological change (Perez, 1983, 1989), so too can shifts in these frameworks affect the development of differentially radical technologies.

References

Akrich, M (1992). The description of technical objects. In: *Shaping Technology: Building Society: Studies in Sociotechnical Change*, Bijker W and Law J (eds.), pp. Cambridge: MIT Press.

Akrich, M and B Latour (1992). A summary of a convenient vocabulary for the semiotics of human and nonhuman assemblies. In *Shaping Technology — Building Society: Studies in Sociotechnical Change*, W Bijker and J Law (eds.), Cambridge: MIT Press.

Alexander, K and PJ Clarkson (2000). Good design practice for medical devices and equipment, Part I: A review of current literature. *Journal of Medical Engineering and Technology*, 24(1), 5–13.

Allen TJ (1966). Studies of the problem-solving process in engineering design. *IEEE Transactions on Engineering Management* 13(2), 72–83.

Arrow KJ (1962). Economic welfare and the allocation of resources for invention. In: *The Rate and Direction of Inventive Activity*, pp. 609–625. Princeton: Princeton University Press.

Baldwin, CY, C Hienerth and E von Hippel. How user innovations become commercial products: A theoretical investigation and case study. *Research Policy* 35(9), 1291–1313.

Berger, AL (1975). Factors Influencing the Locus of Innovation Leading to Scientific Instrument and Plastics Innovation. SM Thesis. Cambridge: MIT Sloan School of Management.

Biemens, WG (1991). User and third party involvement in developing medical equipment innovations. *Technovation* 11, 163–182.

Bijker, WE and T Pinch (1987). The social construction of facts and artefacts: Or how the sociology of science and the sociology of technology might benefit each other. In *The Social Construction of Technological Systems: New Directions in the Sociology and History of Technology*, WE Bijker, TP Hughes and T Pinch (eds.), Cambridge: MIT.

BIS (2009). *Digital Britain*. London: Department of Business, Innovation and Skills.

Brand, G (1972). *The Industrial Buying Decision*. John Wiley and Sons: New York. (Cited in von Hippel, E (1978). A customer-active paradigm for industrial product idea generation. *Research Policy*, 7(3), 240–626.

Braun, V and C Herstatt (2008). The Freedom Fighters: How incumbent corporations are attempting to control user innovation. *International Journal of Innovation Management*, 12(3), 543–572.

Bruland, K and D Mowery (2007). Innovation through time. In: *The Oxford Handbook of Innovation*, J Fagerberg, D Mowery and R Nelson (eds.), Oxford: Oxford University Press.

Caminer, DT, JB Aris, PM Hermon and FF Land (1996). *User-Driven Innovation: The World's First Business Computer*. London: McGraw Hill.

Chatterji, A, KR Fabrizio, W Mitchell and KA Schulman (2008). Physician-industry co-operation in the medical device industry. *Health Matters*, 27(6), 1532–1543.

Cockburn, C and S Ormod (1993). *Gender and Technology in the Making*. London: Sage.

Dahlander, L and MG Magnusson (2005). Relationships between open source software companies and communities: Observations from nordic firms. *Research Policy*, 3(4), 481–493.

DCTI (2007). Innovation Denmark 2007–2010. Danish Agency for Science, Technology and Innovation. Ministry for Science, Technology and Innovation.

DiBona, C, S Ockman and M Stone (1999). *Open Sources: Voices From the Revolution*. Cambridge: O'Reilly.

DiBona, C, D Cooper and M Stone (2006). *Open Sources 2.0: The Continuing Evolution*. Cambridge: O'Reilly.

DIUS (2008). *Innovation Nation*. Department for Innovation, Universities and Skills.

Dosi, G (1982). Technological paradigms and technological trajectories. *Research Policy*, 11, 147–162.

DTI (2004). Global Watch Mission Report: Innovation Through People-Centred Design — Lessons from the USA. Department for Trade and Industry.

Fagerberg, J and B Verspagen (2009). Innovation studies: The emerging structure of a scientific field. *Research Policy*, 38, 219–233.

Flowers, S (2008). Harnessing the Hackers: The emergence and exploitation of outlaw innovation. *Research Policy*, 37(2), 177–193.

Flowers, S and F Henwood (2008). Editorial. *International Journal of Innovation Management Special Issue on User Innovation*, 3(3), v–x.

Franke, N and F Piller (2004). Value creation by Toolkits for user innovation and design: The case of the watch market. *Journal of Product Innovation Management*, 21(6), 410–415.

Franke, N and S Shah (2003). How communities support innovative activities: An explanation of assistance and sharing among end-users. *Research Policy*, 32(1), 157–178.

Franke, N and E von Hippel (2003). Satisfying heterogeneous user needs via innovation toolkits: The case of apache security software. *Research Policy* 32(7), 1199–1215.

Franke, N, E von Hippel and M Schreier. Finding commercially attractive user innovations: A test of lead-user theory. *Journal of Product Innovation Management*, 23, 301–315.

Freeman, C (1968). Chemical process plant: Innovation and the world market. *National Institute Economic Review*, 45, 29–57.

Füller, J, M Bartl, H Ernst and H Mühlbacher (2006). Community based innovation: How to integrate members of virtual communities into new product development. *Electronic Commerce Research*, 6(2), 57–73.

Fuller, J, G Jawecki and H Muhlbacher (2007). Innovation creation by online basketball communities. *Journal of Business Research* 60(1), 60–71.

Gales, M (2006). J. Crayton Pruitt Sr.: Surgeon, Inventor & Biomedical Engineer.

Gassman, O, P Sandmeier and CH Wecht (2006). Extreme customer innovation in the front-end: learning from a new software paradigm. *International Journal of Technology Management*, 33(1), 46–66.

Godin B, 2006. The linear model of innovation: The historical construction of an analytical framework. *Science, Technology and Human Values*, 31, 639–667.

Harhoff, D, J Henkel and E von Hippel (2003). Profiting from voluntary information spillovers: How users benefit from freely revealing their innovations. *Research Policy*, 32(10), 1753–1769.

Henkel, J (2006). Selective revealing in open innovation processes: The case of embedded linux. *Research Policy*, 35, 953–969.

Hienerth, C (2006). The commercialization of user innovations: The development of the rodeo kayaking industry. *R&D Management*, 36(3). 273–294

Jenkins, H (2006a). *Fans, Bloggers and Gamers: Exploring Participatory Culture*. New York: New York University Press.

Jenkins, H (2006b). *Convergence Culture: Where Old and New Media Collide*. New York: New York University Press.

Jeppesen, LB (2005). User toolkits for innovation: Consumers support each other. *Journal of Product Innovation Management*, 22(4), 347–363.

Jeppesen, LB and L Frederiksen (2006). Why do users contribute to firm-hosted user communities? The case of computer-controlled music instruments. *ORGAN SCI*, 17, 45–63.

Jeppesen, LB and MJ Molin (2003). Consumers as co-developers: Learning and innovation outside the firm. *Technology Analysis & Strategic Management*, 15(3), 363–384.

Kent, SL (2001). *The Ultimate History of Video Games: From Pong to Pokemon — The Story Behind the Craze that Touched Our Lives and Changed the World*. Roseville, CA: Prima Publishing.

Kline, R and T Pinch (1996). Users as agents of technological change: The social construction of the automobile in the rural united states. *Technology and Culture*, 37, 763–795.

Knight, KE (1963). A Study of Technological Innovation: The Evolution of Digital Computers. PhD Thesis. Pittsburgh: Carnegie Mellon University.

Lakhani, K and E von Hippel (2003). How open source software works: "Free" user-to-user assistance. *Research Policy*, 32(6), 923–943.

Lettl, C (2007). User involvement competence for radical innovation. *Journal of Engineering and Technology Management*, 24(1–2), 53–75.

Lettl, C and HG Gemuenden (2005). The entrepreneurial role of innovative users. *Journal of Business and Industrial Marketing*, 20(7), 339–346.

Lettl, C, C Herstatt and HG Gemuenden (2006a). Users' contributions to radical innovation: Evidence from four cases in the field of medical equipment technology. *R&D Management*, 36(3), 251–272.

Lettl, C, C Herstatt and HG Gemuenden (2006b). Learning from users for radical innovation. *International Journal of Technology Management*, 33(1), 25–45.

Lüthje, C (2003). Characteristics of innovating users in a consumer goods field. *Technovation*, 23, 245–267.

Luthje, C and N Franke (2003). Bottleneck or Booster of Innovations? A study on the innovation activities of retailers in a consumer goods setting. In: *Proc. of the 2nd World Congress on Mass Customisation and Personalisation*. 6–8 October 2003.

Lüthje, C, C Herstatt and E von Hippel (2005). User-innovators and "local" information: The case of mountain biking. *Research Policy*, 34(6), 951–965.

Marples, DL (1961). The decisions of engineering design. *IRE Transactions on Engineering Management*, 55-71.

Meadows, D (1969). Estimate accuracy and project selection models in industrial research. *Industrial Management Review*.

Morrison, PD, JH Roberts and E von Hippel (2006). Determinants of user innovation and innovation sharing in a local market. *Management Science*, 46(12), 1513–1527.

Mowery, D and N Rosenberg (1979). The Influence of Market Demand Upon Innovation: A critical review of some empirical studies. *Research Policy*, 8, 102–153.

Nelson, RR (1959). The simple economics of basic scientific research. *Journal of Political Economy*, 67, 297–306.

NCM (2006). Understanding User-Driven Innovation. TeamNord 2006:522. Copenhagen: Nordic Council of Ministers.

Ogawa, S (1997). Does sticky information affect the locus of innovation? Evidence from the Japanese convenience-store industry. *Research Policy*, 26, 777–779.

Oudshoorn, N, E Rommes and M Stienstra (2004). Configuring the user as everybody: Gender and cultures of design in information and communication technologies. *Science, Technology and Human Values*, 29, 30–64.

Ourshoorn, N, M Brouns and E von Oost (2005). Diversity and agency in the design and user of medical video-communication technologies. In: *Inside the Politics of Technology*. H Harber (ed.), Amsterdam University Press.

Oudshoorn, N and T Pinch (2003). How users and non-users matter. In: *How Users Matter: The Co-Construction of Users and Technology*, N Oudshoorn and T Pinch (eds.), Cambridge: MIT Press.

Oudshoorn, N and T Pinch (2008). User-technology relationships: Some recent developments. In: *Handbook of Science and Technology Studies*, EJ Hackett, O Amsterdamska, M Lynch and J Wajcman (eds.), Cambridge: MIT Press.

Penning, C (1998). *Bike History.* Bielefeld: Delius & Klasing, Bielefeld.

Peplow, ME (1960). Design Acceptance. In: *The Design Method*, SA Gregory (ed.), London: Butterworth.

Perens, B (1988). *The Open Source Definition.* http://perens.com/policy/open-source/ [31 June 2009].

Perez, C (1983). Structural change and the assimilation of new technologies into the economic and social system. *Futures* 15(5), 357–375.

Perez, C (1989). Technical change, competitive restructuring, and institutional reform in developing countries. World Bank Strategic Planning and Review. Discussion Paper 4. Washington: World Bank.

Prugl, R and M Schreir (2006). Learning from leading-edge customers at The Sims: Opening up the innovation process using toolkits. *R&D Management*, 36(3), 237–250.

Raymond, ES (1999). *The Cathedral & the Bazaar: Musings on Linux and Open Source by an Accidental Revolutionary*. Cambridge: O'Reilly.

Readman, J and A Grantham (2006). Shopping for buyers of product development expertise: How video-games developers stay ahead. *European Management Journal*, 24(4), 256–259.

Riggs, W and E von Hippel (1994). Incentives to innovate and the sources of innovation: The case of scientific instruments. *Research Policy* 23, 459–469.

Robinson, P, CW Farris and Y Wind (1967). *Industrial Buying and Creative Marketing*. Boston: Allyn and Bacon. [Cited in: von Hippel, E (1978). A Customer-Active Paradigm for Industrial Product Idea Generation. *Research Policy*, 7(3), (July), 240–266.]

Rosenberg, N (1982). *Inside the Black Box: Technology and Economics*. Cambridge: Cambridge University Press.

Rothwell, R, C Freeman, A Horsely, VTP Jervis, AB Roberston and J Townsned, (1974). SAPPHO updated: Project SAPPHO phase II. *Research Policy*, 3(3), 258–291.

Schilling, M (2003). Technological leapfrogging: Lessons from the US video game console industry. *California Management Review*, 45(3), 6–32.

Schmookler, J (1962). Economic sources of inventive activity. *The Journal of Economic History*, 2(1), 19.

Schmookler, J (1966). *Invention and Economic Growth*. Cambridge: Harvard University Press.

Schrier, M and R Prugl (2007). Extending lead user theory: Antecedents and consequences of consumers' lead userness. *Journal of Product Innovation Management*, 25(4), 331–346.

Shah, SK (2000). Sources and Patterns of Innovation in a Consumer Products Field: Innovations in Sporting Equipment. *MIT Sloan School of Management Working Paper No. 4105*.

Shah, SK (2003). Community-Based Innovation & Product Development: Findings From Open Source Software and Consumer Sporting Goods. Unpublished Ph.D. thesis, Cambridge, MA: Sloan School of Management, Massachusetts Institute of Technology.

Shaw, B (1985). The role of the interaction between the user and the manufacturer in medical equipment innovation. *R&D Management*, 15(4), 283–292.

Shimshoni, D (1966). Aspects of Scientific Entrepreneurship. PhD dissertation. Cambridge: Harvard University.

Silverstone, R, E Hirsch and D Morley D (eds) (1992). *Consuming Technologies: Media and Information in Domestic Spaces*. London: Routledge.

Taylor, CT and ZA Silberston (1973). *The Economic Impact of the Patent System*. Cambridge: Cambridge University Press.

Turner, F (2006). *From Counterculture to Cyberculture: Stewart Brand, the Whole Earth Network, and the Rise of Digital Utopianism*. Chicago: University of Chicago Press.

Ujjual, V (2009). User-Led Innovation: A Systematic Empirical Review and Synthesis of the Literature. Working Paper. Falmer: SPRU, University of Sussex.

Urban, GL and E von Hippel (1988). Lead user analyses for the development of new industrial products. *Management Science* 34(5), 569–582.

Utterback, JM (1971). The process of innovation: A study of the origination and development of ideas for new scientific instruments. *IEEE Transactions on Engineering Management*, 18(4), 124–131.

Van Kammen, J (2005). Who represents the users? Critical encounters between Women's health advocates and scientists in contraceptive R&D. In: *How Users Matter: The Co-Construction of Users and Technology*, N Ourdshoorn and T Pinch (eds.), Cambridge: MIT Press.

von Hippel, E (1976). The dominant role of users in the scientific instrument innovation process. *Research Policy*, 5(3), 212–239.

von Hippel, E (1978). A customer-active paradigm for industrial product idea generation. *Research Policy*, 7(3), 240–266.

von Hippel, E (1982). Appropriability of innovation benefit as a predictor of the source of innovation. *Research Policy* 11(2), 95–115.

von Hippel, E (2001a). Innovation by user communities: Learning from open source software. *Sloan Management Review*, July.

von Hippel, E (2001b). Perspective: User toolkits for innovation. *The Journal of Product Innovation Management*, 18, 247–257.

von Hippel, E (2005). *Democratizing Innovation*. Cambridge: MIT Press.

von Hippel, E and SN Finkelstein (1978). Analysis of innovation in automated clinical chemistry analysers. *Science and Public Policy* 6(1), 24–37.

von Hippel, E and G von Krogh (2003). Open source software and the "Private-Collective" innovation model: Issues for organization science. *Organization Science*, 14(2), 208–222.

von Krogh, G and E von Hippel (2006). The high promise of research on open source software. *Management Science* 52(2), 975–983.

von Tunzelmann, N, F Malerba, P Nightingale and S Metcalfe (2008). Technological Paradigs — past, present and future. *Industrial and Corporate Change*, 17(3), 476–484.

Voss, CA (1985). The role of users in the development of applications software. *Journal of Product Innovation Management*, 2, 113–121.

Woolgar, S (1991). Configuring the User: The Case of Usability Trials. In: *A Sociology of Monsters: Essays on Power, Technology and Domination* (Law J ed.). London: Routledge.

THE DYNAMICS OF USER INNOVATION: DRIVERS AND IMPEDIMENTS OF INNOVATION ACTIVITIES

CHRISTINA RAASCH, CORNELIUS HERSTATT and PHILLIP LOCK

Hamburg University of Technology, Germany

Introduction

It is widely accepted today that users or user networks are often a major source of innovation and have even proven to be the principal driving force of many innovations in different industries (Lettl *et al.*, 2006). For this reason, more and more firms try to identify avenues to systematically involve users into their new product development.

Eric von Hippel and his scholars in the early 1980s were the first to systematically document that users, rather than manufacturers, are often the main source of innovation (von Hippel, 1977, 1988). Nevertheless, even Adam Smith reported in "The Wealth of Nations" that many innovations in machinery were actually developed by workers wishing to make better use of production facilities (Smith, 1776; cited in von Hippel, 2005, p. 21).

Despite the growing interest in user-driven or user-centred innovation, both in academia and industry, the various mechanisms and principles of the "democratisation of innovation" (von Hippel, 2005) are not yet fully understood and only few researchers have been looking more closely at barriers, which prevent users from becoming active innovators (e.g., Braun and Herstatt, 2007). Further, the drivers and impediments affecting the evolvement of user innovation-related activities over time have only recently become a focus of analysis (Baldwin *et al.*, 2006; Braun, 2007).

With our study, we aim to investigate user innovation over time and contribute to the extension of the existing model of user-driven innovation to a more dynamic setting. For this purpose, we analyse the evolution of user innovation activity in the field of sports equipment, drawing on the case of a high-performance sailing boat called *Moth*. By this choice, we aim to facilitate comparisons and contrast findings to previous research results in the field of sport equipment. We analyse innovation activities in this field over several decades based on secondary data, in-depth interviews and results from a survey we conducted in 2007. We find that the level of user innovation does not follow a unidirectional trend, but rather rises and falls depending on a number of contextual factors. This suggests that, given a stimulating setting, user innovation can be sustained over periods of time.

Our paper is organised as follows: We first present a review of the literature. The section 'Overview of the Moth class' provides a brief description of the characteristics of the Moth boat class that, from our perspective, make it an ideal field for this research. In particular, we point out high levels of successful user innovation activities and the long history of the class itself. Subsequently, we present the results from our empirical survey among Moth sailors (in the section on 'Empirical findings on user innovation activity'). We identify different types of users belonging to the community and their preferences for innovation vs. standardisation, information sharing behaviour and preferred ways of communication. In the section on 'The dynamics of the innovative activity of users', we investigate the dynamics of user innovation activities in the International Moth Class over the last 80 years. We conclude this paper with a discussion of our findings and implications for industry practice.

Literature Review

According to the 'manufacturer-active paradigm' (MAP) (von Hippel, 1978), the entire sequence of activities requisite for the launch of an innovation solely belongs to manufacturers' responsibility. This traditional understanding of the NPD process leaves to users the role of "passive acceptors of an innovation" (Rogers, 2003, p. 180).

The past three decades of academic research have shown that users, contrary to the MAP, often take an active role in the innovation process, testing and modifying existing products and even designing new ones themselves. Much research has been devoted to the role of lead users in particular, who were found to be more likely to innovate than users in general (Urban and von Hippel, 1988). Lead users are defined by two characteristics: first, they are at the leading edge of markets, featuring needs that will become prevalent in the marketplace in the future, and, second, they benefit substantially from a solution to these needs (Morrison et al., 2004). In some cases, lead users may actually be "the more suitable innovators" (Braun, 2007, p. 17) because they are cognizant of the specific aspects in which a standardised commercial product falls short of their needs. User need information is intangible and sticky (von Hippel, 1994), rendering it difficult for manufacturers to obtain. Moreover, even if manufacturers were somehow given all the required information on user needs, need heterogeneity can often not be satisfied by standardised, mass-produced solutions (Franke and von Hippel, 2003; Jeppesen, 2005).

'Customer-driven innovation' was mostly analysed at an industry level, proceeding on the assumption that some industries are more suited to user innovation than others. Innovative activities undertaken by lead users have been studied, e.g., for petroleum processing (Enos, 1962), medical equipment (Shaw, 1985; Biemans, 1991; Lettl et al., 2006), fastening equipment (Herstatt, 1991), scientific instruments (Riggs and von Hippel, 1994), library information systems (Morrison et al., 2000), drug therapies (DeMonaco et al., 2005), sports equipment (Shah, 2000; Franke and Shah, 2003; Lüthje, 2004; Lüthje et al., 2005; Füller et al., 2005; Tietz et al., 2005), and open-source software (von Krogh and von Hippel, 2006, provide an overview).

Lately, however, research attention has shifted to the role of users across the different phases of the innovation process (Lettl et al., 2006; Hienerth, 2006; Baldwin et al., 2006). A more dynamic understanding has emerged, indicating that user innovation-related activities not only differ among industries, but also within industries over time. These findings indicate that user activity increases after some users discovered a set of new design possibilities; successful new solutions eventually lead to rising demand for the new design and thus commercial offerings by user-manufacturers; as the unexploited potential of the design diminishes and users therefore reduce their efforts, the market is taken over by large firms selling standardised, mass-produced solutions (Baldwin et al., 2006).

Some evidence suggests that a number of other factors influence the level of user innovation activities over time, possibly disrupting the evolvement according to the four-stage process delineated above (Baldwin et al., 2006, pp. 1307–1308). Braun and Herstatt (2007) analyse how the erection of legal, technological, economic and social barriers adversely affected the extent of user innovation in the field of seed-breeding. Braun (2007) predicts that "[t]here will be a constant struggle between encouraging and discouraging effects which will allow user-innovation in some areas at some occasions to flourish, while at others to vanish" (p. 172). Still, many factors relating to the dynamics of user innovation are not fully understood or explored yet. With this work we aim to illuminate some of these factors based on an empirical case example, thereby suggesting avenues for further research on the evolvement of user innovation over time.

Overview of the Moth Class

Sailing is different from some other sports environments hitherto described in the literature in that it consists of many different boat classes with their own communities, rules and cultures. Most sailing dinghies have to conform to strict class regulations regarding technical specifications, measurements and materials used. These rules often remain unchanged for many years or even decades, until some adaptations are made to accommodate technological or other changes. The most extreme among these classes are so-called one-design classes, which consist of virtually identical boats, a principal rationale being that users shall compete based on sailing skill alone. Moreover, design stability allows economical runs of production. At the other end of the spectrum, the so-called development classes allow for considerable variety and regard experiments undertaken by the community as a source of progress. One class from the latter group is the International Moth Class, the focus of our research.

An International Moth is a small performance sailing dinghy with a length of approximately 3.4 m (11 feet) that carries one person. It is estimated that there are less than 1000 Moth sailors world-wide, although this number is currently increasing rapidly. The first Moth was built in southern Australia in 1928 (Fig. 1). Since then, its open rules have rendered the Moth class one of the most innovative classes and a source of innovative features later taken up by other classes. Major instances of the Moth's pioneering role are hiking wings that extend the deck sideways, thus enabling the sailor to move his bodyweight further outwards, and T-shaped rudders that create lift. A recent innovation is the so-called hydrofoil technology, horizontal

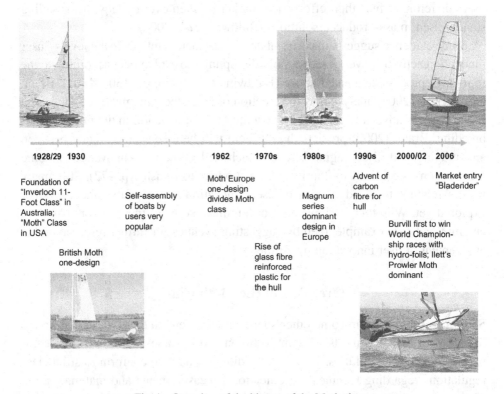

Fig. 1. Overview of the history of the Moth class.

wings on the end of the rudder and the centreboard, which lift the hull completely out of the water. Thereby, friction is reduced, allowing the Moth to reach speeds of up to 55 km/h. In the wake of the introduction of hydrofoils, interest in the Moth class has increased considerably and market demand is rising.

While performance-enhancing modifications undertaken by users on existing boats are the rule in the Moth class, boats are built by three groups: Open rules encourage sailors to build their boats themselves, using existing building kits or plans or even creating entirely new designs. Still, professionally-built boats nowadays make up the majority of new boats. Moth manufacturers were — until recently — all small companies, producing boats according to users' specific requirements. Their owners are mostly current or former Moth-sailors. Lately, one large manufacturer, the Australian-based Bladerider International, has entered the market with production facilities for 2000 boats per annum located in China (Bladerider Int., 2007). Bladerider currently offers the most successful design, accounting for three out of the best five boats in the 2007 World Championship. With this design, Bladerider aims to build a new one-design class as a Moth spin-off and an alternative to community members wishing to evade the constant demands of boat improvement.

For three reasons, the environment just described offers a good test case industry to our research on the dynamics of user innovation: First, it features high user activity — one interview partner stated that "in the Moth class, all innovation is user innovation". Second, its relatively long history favours longitudinal research. And third, its development can shed light on several factors influencing user activity that are not yet well understood. Before we continue to examine the dynamics of user innovation, however, survey results will be presented in the following section to describe user characteristics and innovation activities in the Moth class.

Empirical Findings on User Innovation Activity

A survey among Moth sailors was conducted during the 2007 World Championship. A questionnaire was prepared based on seven expert interviews, each of approximately 2 hours length, supplemented by further data collection via e-mail, and was improved during two rounds of pre-tests. Fifty three sailors participated in the survey.

The majority of questions use a 7-level Likert scale, where 1 stands for "I do not agree at all" and 7 for "I agree completely". Apart from descriptive analyses, we use the Kruskal-Wallis test as a non-parametric test of the equality of population means. It was chosen because the normality assumption does not hold for our sample. In the text, α stands for the significance level at which the null hypothesis of equal population means can be rejected based on the H statistic.

Characteristics of user innovators

Lüthje *et al.* (2005) find that user innovators have been involved in their sports (mountain-biking) for a significantly longer time than non-innovators (similarly Franke and Shah, 2003). Our data confirm this finding: On average, especially innovative users (i.e., users who perceive themselves to be contributing more ideas than the rest of the community) have been active in the class for an additional 6.7 years compared to the average. Sailors with long class affiliation tend to be more successful in races ($\alpha < 0.2$), a fact that points to the importance of experience in sailing technique and improvement of materials. Success in competition is closely related to the frequency with which users see their modifications copied by other community members or taken up by manufacturers (all $\alpha \leq 0.03$).

Innovative users believe in the universal presence of opportunities of improvement, and the majority of them have even built a Moth boat themselves (all $\alpha \leq 0.03$). This corroborates the estimation voiced by interview partners that, although self-designed boats nowadays have little chance of winning in major competitions (see the section on "The dynamics of the innovative activity of users"), they are still an important source of new ideas.

Sailing tinkerers and tinkering sailors

A proverbial truism among Moth sailors is that the class has always been made up of sailing tinkerers and tinkering sailors. In other words, activities by users aimed at improving their boats are almost universal. This estimation was confirmed during our interviews with respect to class history and also manifests itself in the survey results. Being asked for their degree of agreement to the statement "To me working on my boat is an integral part of Moth sailing", 70% strongly agreed (level 6 or 7), 13% of the participants strongly disagreed (1 or 2), and 17% did not have strong feelings in either direction (3–5). The high level of user activity in the Moth class supports findings from earlier studies in the fields of sports equipment (Shah, 2000; Franke and Shah, 2003; Lüthje, 2004; Tietz *et al.*, 2005).

The dictum also implies that the preference for expending effort on boat improvement is stronger with some dedicated tinkerers, while others primarily want to pursue their hobby, and, as a pre-requisite, also need to do some work on their boats (similarly Hienerth, 2006). For the latter group, standardisation is often welcome, presumably because it allows them to reduce their effort without losing out in races. Accordingly, 50% of the survey participants are strongly averse (1–2) to buying a one-design Moth, which would considerably lessen degrees of freedom in modification activities, while 21% would seriously consider doing so (6–7). Besides setting a lower value on tinkering, the group preferring a standardised design devotes considerably less time and money to boat improvement and is less involved in information exchange with the community via the Internet (all $\alpha \leq 0.05$).

This coexistence of users with different preferences for innovation vs. standardisation is confirmed by a number of episodes in Moth history. At different times, the preferences of some groups are known to have leant towards higher standardisation to curb the constant pressure to improve material; tightened rules or the secession of the *Moth Europe* as a one-design standardised spin-off class were some of the consequences. Other sailors refused to participate in standardisation tendencies and even continued tinkering in violation of tightened class rules, knowing that their designs would not be allowed in competitions.

Knowledge-sharing

In line with previous empirical analyses in sports communities (e.g., Shah, 2000, 2004; Franke and Shah, 2003; Lüthje *et al.*, 2005; Hienerth, 2006), we find that knowledge is mostly shared freely within the Moth community (Fig. 2). While, in the light of prior results, it may not be surprising that users are prepared to share approximately 94% of their information with sailing friends or colleagues, the fact that they would share 80% of their knowledge with direct competitors seems noteworthy. Earlier findings (Franke and Shah, 2003) suggest that competition reduces sharing and assisting behaviour among users to a larger extent.

We also find that the percentage of knowledge shared with manufacturers substantial (cf. Füller *et al.*, 2005), but significantly lower than the proportion shared with individual community members. Keeping in mind that small manufacturers are mostly also users and have been for many years or even decades, this difference in communication behaviour seems hard to explain. We assume the distinction

	$\bar{x}$	Median	s
What percentage of your knowledge would you share with the Moth Class in general?	80.7%	100.0%	26.3
What percentage of your knowledge would you share with friends?	93.6%	100.0%	18.5
What percentage of your knowledge would you share with sailors you compete with directly?	81.5%	100.0%	25.4
What percentage of your knowledge would you share with manufacturers?	72.4%	90.0%	34.5
What percentage of your knowledge would you share with sailors who do not share their information?	41.6%	50.0%	37.8

($N = 53$)

Fig. 2. Free revealing in the Moth class.

to be attributable in some measure to the resentment one particular manufacturer, Bladerider Int., has lately attracted for "hijacking" (O'Mahony, 2003) ideas developed by the community over many years.[1]

With an average evaluation of 4.8 on a scale from 1 to 7, the majority of respondents agree that the Moth class is somewhat more competitive than other boat classes, and 30% consider it to be much more competitive (level 6 or 7). Somewhat contrary to our expectations (Harhoff et al., 2003), there is no evidence that this group of users shares less knowledge with other community members. The only significant difference relates to information sharing with direct competitors, to whom a still impressive 74% of knowledge — compared to the overall average of 81% — is revealed. These results support earlier results by Franke and Shah (2003) who found that information sharing declines, albeit not strongly, with the degree of competitiveness of the community.[2]

According to von Hippel (2002, p. 5), "participants [in sports communities] tend to freely reveal their innovations to all participants, including free riders". Our data corroborate this statement in general. We find that free-riding does occur in the Moth class. Sailors are confident that the class mostly recognises this behaviour as such after some time, but no specific sanctions are attached. Still, the proportion of knowledge that sailors are willing to share with free-riders is significantly lower than what they would share in general (approximately 40% instead of 81%). This corresponds to a perception of fairness in generalised exchange settings (Takahashi, 2000) that is also prevalent in other communities; i.e., in open source software, where the basic principle for participation is that "there is no limit on what one can take from the commons, but one is expected at some time, to contribute back to the commons to the best of one's abilities" (O'Mahony, 2003, p. 13).

With regard to the motives for sharing information this freely, we find strong agreement to the intrinsic and extrinsic motives usually attested to user innovators in empirical studies (e.g., Shah, 2000; Hars and Ou, 2002; Franke and Shah, 2003; Füller et al., 2005). The strongest drivers, according to our survey, are the expectation

---

[1] In the year 2002, the most successful foiling Moth to date was designed by John Ilett's company Fastacraft, a small Australian manufacturer. Patent protection was not applied for, not only due to high costs and effort but also because patentability seemed uncertain since many users had been involved in the design and improvement process over the years. Given its lack of patent protection, the design was taken up by Bladerider Int. in 2005 and subsequently fine-tuned during further professional experiments. Bladerider has recently applied for patent protection for some components, causing considerable discussion within the community as to the role of patents in development classes.

[2] It should be noted, however, that information sharing in highly competitive environments — in the realm of sailing e.g., Olympic one-design classes or professional sailing as in the America's Cup — seems considerably more limited.

	$\bar{x}$	Median	s
Usage of the Internet to gather information (a):	4.9	5	1
Usage of regattas/private contacts togather information (b):	5.9	6.5	1.2
Usage of the Internet to pass on information (b):	3.5	3.5	2
Usage of regatta/private contacts to pass on information (b):	5.6	6	1.4

(a): aggregated from 3 individual questions, 1 = I do not agree at all, 7 = I agree completely
(b): aggregated from 2 individual questions, 1 = I do not agree at all, 7 = I agree completely

($N = 53$)

Fig. 3. Media of information exchange.

of reciprocation, enjoyment derived from helping others and the awareness that progress in a community, which advances largely by user innovations, relies on the exchange of information.

In terms of the preferred medium of exchange, previous research indicates that "participants in sports innovation networks tend to interact by physically travelling to favourite sports sites and to contests designed for their sport" (von Hippel, 2002, p. 5; cf. Tietz *et al.*, 2005). In the Moth community, face-to-face communication, for example during regattas, is indeed favoured over exchange via the Internet (including several Moth websites, private blogs and forums). This is particularly true with regard to users' inclination to share their knowledge, whereas information gathering is strongly supported by digital means of communication as well as personal meetings (Fig. 3). As to the reasons favouring offline communication, convenience was pre-eminent: Sketches of envisaged changes and possibly the display of physical components were considered the easiest and fastest way of communicating ideas. The wish to ascertain and possibly restrict the group of information recipients played a subordinate role.

Differences in community culture

The fact that face-to-face communication is still the preferred means for the exchange of ideas combined with the fact that international regattas take place once or twice each year implies that communication can be expected to be more extensive within national sub-communities than across borders. This information stickiness (Lüthje *et al.*, 2005) may even result in national sub-communities with somewhat different cultures. Our data give tentative support of this hypothesis, despite low overall numbers of respondents from each country.

Australia, Great Britain and Germany are home of the largest national class associations. For example, 42 out of 53 respondents hail from these countries, with 14, 15 and 13 sailors in our sample coming from each respective country. The Australian community use the latest boats, score best in races, and believe that they contribute most of the new ideas. German users, by contrast, on average have considerably older boats, mostly score in the bottom third, and rarely encounter interest in their modifications (all $\alpha \leq 0.05$). British users reside in between the two.

Although there is no significant difference in the perception of the competitiveness of Moth sailing, Australians share significantly less information than Germans (72% on average vs. 90%). Germans were more inclined to desire to help others with the information they shared and utterly rejected the statement that they felt mostly resigned when others copied their ideas; Australians declared an intermediate level of agreement of 4.3 to this statement. British sailors rank between the more competitive Australians and the more community-oriented Germans. Thus, while information sharing is clearly common in all the three sub-communities, the sharing culture does seem to differ considerably. One of the underlying reasons could be that Moth sailing in Australia is more professional in the sense that more sailors derive revenue from sailing, e.g., as user manufacturers or from sponsorships by manufacturers.

The Dynamics of the Innovative Activity of Users

Revolution and adoption

Our empirical observations, particularly regarding the high level of user innovation-related activities and the different preferences for innovation within the Moth class, prompt the analysis of user innovation activity across time. For this purpose, we would need a measure that is fit for longitudinal study over the last 30–50 years and captures user-generated innovations as well as an evaluation of the degree of newness. Unfortunately, no such direct measure of user innovation activities across decades of Moth class history is available. Therefore, we propose the following proxy: Studying the age of the design of the boats winning in national and international championships gives us an idea of the pace of technological progress at the time: If only newly designed boats score highly, this suggests a marked improvement of design, whereas older boats winning races point to a slow-down in progress (cf. Baldwin et al., 2006, who follow a similar approach with regard to tournament scores). We assume that progress in boat design can be traced, at least to a substantial degree, to user activity. This assumption is grounded in the observation that in the Moth class, most innovative activity involves users (see the section on "Overview of the Moth class").

The validity of this proxy is corroborated by an analysis of the breakthrough innovations in the period under consideration (Fig. 4). The analysis is based on

Phase	Innovation	Year of introduction	Year of first success in international championship
1	Hiking wings	1968	1973
	Skiff	1968	1975
2	Narrow skiff	1988	1992
	T-foil rudder	1994	1994
3	(Bi-) Hydro-foils	1999	2004

Fig. 4. Breakthrough innovations in Moth design.

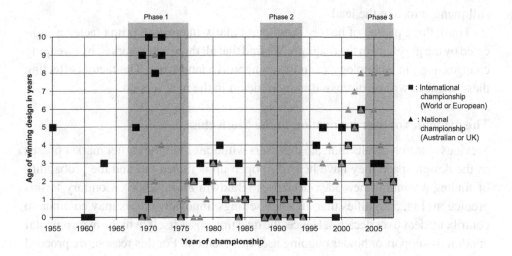

Fig. 5. Pace of design progress in Moth class.

documentation and expert interviews, each from inside and outside the Moth class. The most innovative design steps and their subsequent diffusion within and sometimes beyond the Moth class fall mostly well within the three phases of change we identify using our proxy variable (see below). The hydro-foil technology was initially slow to diffuse due to its extremely difficult handling.

Figure 5 plots the age of the design of the boats winning international and national championships from 1955 to 2007.[3] It prompts three principal conclusions: First, there is no clear-cut trend across time. In particular, if a dominant design emerged, we

---

[3] If no or just one entry is made for a specific year, the championship was taken by a boat that was designed and built entirely by a user. In this case, no exact information on the age of the design is available.

would expect to see boat age lose relevance for success at some stage. While there are phases in which one specific design is particularly strong (e.g., the Magnum series in the early 1980s), we find no evidence of a permanent decline in user activity. Survey results support this tentative finding. Being asked for the level of agreement to the statement "I think there will be less innovation from hobby Moth sailors in the future", 17% strongly supported the statement (6–7), while 47% of respondents strongly disagreed (1–2). Support for the statement was notably low even among strong proponents of a one-design class spin-off.

Second, we find some support of a cyclical pattern in the pace of design progress. Across some stretches of time, only very new designs are sufficiently fast to win in championships, whereas in other periods boats designed several years previously still manage to take the lead.

Third, three phases of history seem particularly interesting prima facie, as indicated by the grey bars in the diagram. We find that all three were marked by sweeping changes, e.g., in technology or in the competitive landscape. The factors effecting these changes will be discussed in more detail in the next section.

The dynamics of user innovation in the Moth class

Previous research results indicate that users withdraw after the earlier market phases as the design space they have helped to open up is mined out and the probability of finding a winning new idea decreases (Baldwin *et al.*, 2006). Contrary to this prediction, Fig. 5 signifies that, in some settings, innovative users may continue to contribute ideas over decades. Hence, the question poses itself, which environmental conditions support or hinder ongoing user involvement. For this reason, we proceed to look at different phases of Moth class history in order to identify driving forces or impediments to user innovation. We propose that five factors influence the level of user activity at any point in time.

Technology complexity

From the research undertaken by Braun (2007) and Hienerth (2006), we assume that user innovation tends to decline as technological complexity increases. Based on our data, this pattern seems also applicable with regard to Moth development.

Figure 6 indicates years in which self-assembled boats, i.e., boats designed and built without the involvement of manufacturers, won in international and national championships. While in Phase 1, during the 1950s and 1960s, at least one international championship was usually taken by such boats, the 1970s were marked by the take-over of manufacturer designs. One reason for this change was the successive replacement of plywood by glass fibre reinforced plastic for the hull of the boat. With its increase in strength and decrease in weight, this new material became a pre-requisite for remaining competitive. However, it restricted the chance to design

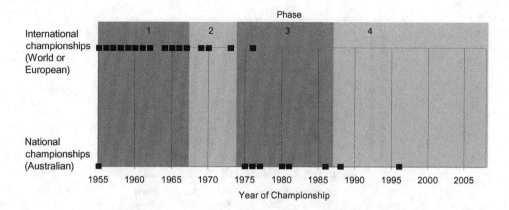

Fig. 6. The success of self-constructed boats in championships.

high-performance hulls to only a handful of users who had access to the technology and required manufacturing equipment. The only victories of self-constructed boats thereafter achieved occurred at the national level in Australia. The Australian community at the time was slow to adopt the new plastic materials (and the narrower hull design these materials enabled), but adhered to broad blunt-bowed scows made from plywood (International Moth Class Association of Australia, 2008).[4]

Phase 4 was marked by the success of even narrower hulls (the so-called narrow skiff design with a width of approximately 0.25 m at water level) and the rise of carbon fibre, which essentially heralded the end for self-assembly in the Moth class. The moulds and ovens necessary to build hulls from carbon became prohibitively expensive for users, unless their profession allowed them to access the requisite equipment (cf. Hienerth, 2006).

However, this step in material technology did not bring user innovation activity to an end. Instead, users focused their attention on other parts of the boat, particularly the rudder and centreboard. As a side-note, the community is currently evaluating a relaxation of class rules to reverse the narrowing of tinkering scope imposed by increasing technological complexity.

Technological maturity

After user attention shifted to the rudder and centreboard in the 1990s, several scattered experiments led to the incubation of the hydrofoil technology (see the section on "Overview of the Moth class"). The stability of a boat "flying" half a

---

[4]Note: For Australian self-assembly, data for 1955 to 1975 is unavailable. Thus, the lack of data points towards the beginning of the period under review does not imply that self-assembly was not successful then.

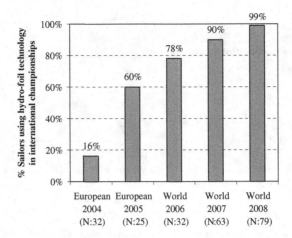

Fig. 7. Percentage of sailors using the hydro-foil technology in international championships.

meter above the water surface, with nothing but wings or "foils" regulating its angle and height, posed serious problems. Users and user manufacturers collaborated and finally came up with several designs around the turn of the millennium. Between 2003 and 2007, boat speed with hydrofoils increased by 30%, and foils are now standard equipment for any sailor wishing to compete effectively in races (Fig. 7).

However, with stability issues now essentially resolved, the fine-tuning of height control yields declining incremental benefits. Our interview partners confirmed that attention has now turned to the rigging (i.e., mast, boom, and the wiring required to attach them to the hull). Users believe that another performance step will result from a better adaptation of the rigging to the new requirements posed by a boat that is not so much sailing as flying.

To summarize, user activity in this field was high when the technology was immature (cf. Ozer, 1999; Hienerth, 2006). The reason for this is clear-cut: When a new idea or technology adds dimensions to the design space of techniques that can be mined for improvements, the chances of users to develop a successful modification are comparatively high. Subsequently "total search effort by user-innovators will tend to decline over time [as] the expected improvement in the design will no longer exceed that person's perceived value of time, effort, and expense" (Baldwin *et al.*, 2006, p. 1299). However, our study points to an extension of the Baldwin *et al.*-model whereby "the collapse of the user-driven experimental sector of the industry" eventually results (Baldwin *et al.*, 2006, p. 1307). Our findings suggests that, when the incremental return to tinkering declines as technology matures, user attention, rather than waning, may shift to new fields of activity, which promise a higher return on expended effort and investment. In this sense, users may be regarded as allocating their efforts across potential fields of activity, maximising the return expected to derive (cf. Dosi *et al.*, 2006). Their behaviour, seen in this light, resembles the behaviour of entrepreneurs deciding in which potential fields of business to invest.

Market structure

Previous research has shown that, as the rate of user innovation in a design space declines, investment in manufacturing equipment with higher capital cost and lower variable cost becomes profitable (Utterback and Abernathy, 1975; Baldwin *et al.*, 2006). Capital-intensive production of a highly standardised design helps to reduce production cost and product price and to increase market size. This implies that large capital-intensive manufacturers can frequently not accommodate users' innovative activities or requests for design modification (examples of mass customisation forming an exception).

Observations from our study confirm that user innovation receives more support from small-scale manufacturers with more flexible production processes than from large suppliers with more capital-intensive production (Braun, 2007). Users who purchase a boat design from one of the small manufacturers can combine components from different suppliers (hull, rudder and centreboard, sail, mast etc.) and require the manufacturer to make additional adaptations specific to their requirements. In contrast, the Bladerider design, the only one produced on a larger scale, is a one-design offering no modification options to users. After-sale changes to the boat remain possible, but may sometimes not be easy to implement. Thus, differences in boat design can be expected to diminish considerably, if Bladerider Int. as a large-scale manufacturer wins market share from small user-manufacturers.

Customer satisfaction

The displacement of the user-driven sector of the industry by large commercial suppliers critically hinges on there being sufficient demand for a standardised commercial solution. In other words, it presupposes that customers are satisfied with the new design they have helped to create, perceiving further search for improvement to promise insufficient benefit (on dissatisfaction with current market solutions as a trigger of user innovation cf. Franke and Shah, 2003; Lüthje *et al.*, 2005; Hienerth, 2006).

Our example from paragraph (2) relating to the different components of boats (hull, rudder, rigging etc.) gives some evidence that user-innovators may consecutively open up new areas of the design space, thus never ceasing their innovative activities, but just shifting their focus. In our case example, at least some users believe that the benefits of further tinkering are likely to surpass the cost. This relates to the empirical observation discussed in the section on "Overview of the Moth class" that preferences for tinkering vary across community members.

Even in cases in which customers are mostly satisfied with the standard they have helped to establish, the level of satisfaction and thus demand may actually shift across time. As either the expected benefits or the costs of innovative activity change for users, the proportion of user innovators changes over time, as the history of Moth sailing illustrates: In the 1930s, the 1960s and 1970s, and again in recent years, standardised solutions (the British Moth, Moth Europe, and Bladerider designs) attracted many sailors. While the British Moth has retained very few enthusiasts today, the Moth Europe from 1962 still attracts a thriving community in many countries. However, both designs are completely outdated, lagging far behind innovative new designs in terms of materials, construction, weight, and thus speed.[5] For this reason, some community members felt increasingly frustrated by the comparative slowness of their boats. Thus, the group of dedicated tinkerers, after initially seeing two thirds of all class members opt for a standard design, subsequently won back followers. In a similar fashion, the modern Bladerider one-design attracts many users with its low maintenance and strong performance. Still, all our interview partners including Bladerider representatives were certain that continued user activity would re-establish its performance lead within a few years, possibly re-attracting Bladerider owners to non-standardised boats.

Barriers to innovation

Innovation barriers erected by manufacturers can put a curb on user innovation activity (Braun, 2007). In our industry example, such barriers were historically absent or negligible (cf. Harhoff *et al.*, 2003; Lüthje *et al.*, 2005). The recent advent of the Bladerider design, however, has introduced restrictions on innovative activity undertaken by users. For example, the company does not sell single components to owners of competing boat designs and has recently applied for patent protection for some components, causing resentment among some community members for the "hijacking" of their ideas (see also footnote 1). To guard against such problems, Füller *et al.* (2005) recommend that "issues regarding intellectual property and gratifications have to be clarified before a company may collaborate with communities" (p. 21).

It should be noted that not only manufacturers, but also users may profit from erecting innovation barriers themselves. In our study, users at different times felt the pressure to expend effort in order not to be left behind by more active user innovators to be excessive and therefore tried to establish some stable standard (e.g., the Moth Europe one-design).

---

[5] A Moth Europe, for instance, weighs twice as much as a modern Moth and is more than four times as wide.

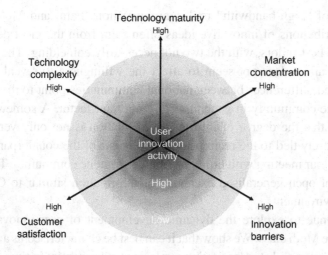

Fig. 8. Factors driving user innovation-related activities across time.

Factors driving the development of user innovation across time

Based on the analysis in this section, we propose that the activity level of user innovators at any point in time is affected simultaneously by five factors, as summarised in Fig. 8.

These factors may jointly produce the cyclical pattern of innovative activity and progress we portrayed in our case example. They interfere with the clear-cut four-phase progression from earliest lead user experiments to standardised commercial solutions, as identified by Baldwin *et al.* (2006). For the example of the Moth sailboat, we rather find a sequence of revolution and adoption, as proposed by Schumpeter (1942), without the lasting emergence of a dominant design.

Summary and Discussion of Findings

Summary of results

With this contribution, we aim to shed light on the factors influencing the level of user innovation-related activities over time. As an industry case study, we chose a class of high-performance sailboat, the so-called International Moth Class, in which a large proportion of innovations can be traced to modifications undertaken by users. After a description of user activities in this boat class based on interviews and survey results, we proceeded to analyse the dynamics of user innovation activities over the last decades.

Survey results support previous findings regarding the high level of user innovation activity and the prevalence of a community sharing culture in the field of sports. We analyse free revealing behaviour and the media used, highlighting the importance of "high-bandwith" oral communication (Faraj and McLure Wasko, 2001). Contributions of innovative ideas often stem from the most dedicated tinkerers or the best sailors, with the two not necessarily coinciding. The competitive position of sailors does not seem to affect the willingness to reveal information freely; instead, differences between national communities point to the competitive culture of the community in general as an important factor. A somewhat surprising result is that the degree of information revelation is not only very high; it is also only loosely tied to the reciprocating behaviour of the counterpart, with free-riding behaviour meeting with little sanctioning from the community. Thus, we find a situation of open generalised exchange of information similar to Open Source Software environments.

We continue to explore the dynamic development of user innovation-related activity in the Moth class. We show that it can best be characterised as a *co-existence* of user innovation and standardisation tendencies at each point in time. Several factors propel the balance to shift in a somewhat cyclical pattern, one being the level of satisfaction of users with the solution they have helped to create. We find that standardised solutions are often unable to emulate the performance improvement effected by innovative users over the years, so that standardised solutions inevitably become outdated. In the words of Raymond (1999) there is an "evolutionary arms race [which the] closed-source world cannot win" (p. 41). Apart from user satisfaction with existing design, we identify the complexity and the maturity of technology, market structure, and barriers to innovation as factors influencing the level of user innovation at each point in time. We find that depending on the characteristics of the environment, users may not cease innovating after some initial period, as suggested by previous studies (Hienerth, 2006; Baldwin *et al.*, 2006), but shift their efforts to new fields, which promise a higher return on expended effort and investment.

Generality of findings

The successive re-allocation of user efforts to different fields of activity, which we observed in the Moth class of sailboats, can be traced to two reasons. First, barriers to user innovation in the form of technological, legal or regulatory hurdles may rise or be erected, blocking some area of the design space, causing users to re-focus on other areas. From the user perspective, such changes are mostly exogenous. And second, the successive mining out of one field renders other fields of activity with their expected returns to user involvement more attractive.

Following Baldwin *et al.* (2006, p. 1296, and references cited therein), we regard the design space in our case to be the entire Moth sailboat. What we find is that the design space is consecutively expanded or "augmented" (Baldwin and Clark, 2000) by radical user ideas, e.g., hiking wings or hydrofoils. Such disruptive technologies allow users to move beyond the design space, which hitherto seemed finite and clearly circumscribed. After the newly-discovered subset of the design space, e.g., the hydrofoil technology, is mined out, users turn to other subsets, successively expanding and mining-out new areas of the design space. Thus, based on our case study evidence, no mining-out of the entire space seems to take place.

What does this imply for the generality of our findings? Clearly, users can only 'move on' to new fields of activity, if the borders of the design space are not inflexible, i.e., if the design space is not finite. Unless augmentation is possible, we should therefore expect user innovation to decline and eventually cease. This pattern would only be disturbed by exogenous changes to the conditions for innovation, e.g., to the regulatory framework, materials etc. The assumption of fading user innovation activity in finite design spaces is supported by earlier research in the rodeo kayak industry (Baldwin *et al.*, 2006). Evidence from more strictly regulated classes of sailboats, e.g., one-design classes, where modifications quickly hit the regulatory boundary of the design space, points in the same direction. In such environments users cease to innovate to any substantial extent when the design space has been fully explored. 'Out-of-the-design-space' ideas would yield little benefit inasmuch as their implementation results in exclusion from the class. This suggests that, absent exogenous shocks, sustained user activity can be found in expandable design spaces rather than in tightly limited ones.

What seems noteworthy in this context is that a design space that seems finite at any point in time, e.g., a rather small and not overly complex sailboat, may later be expanded by active users who do not accept the limitations of the design space they work in. In this light, our findings may be applicable to many areas which currently seem finite but may be proven by users to be augmentable.

Implications for managerial practice

Based on our case example, we believe a further theoretical and empirical evaluation of the five drivers of user innovation we identified to be a rewarding undertaking. A better understanding of these factors will be of value to practitioners trying to gauge the dynamics of current user activity or the potential success of a future involvement of users in the NPD process. If manufacturers manage to shape contextual factors so as to stimulate protracted user activity, they may benefit greatly from users consecutively opening up and exploring new areas of the design space. Our findings suggest that, in order to reap these benefits, they should take care not to put tight restrictions on the design space they wish users to explore. Limitations, which hinder the expansion of the design space, are likely to cause users to discontinue their efforts or to move on to other design spaces with higher expected returns to their innovation activities.

References

Baldwin, CY and KB Clark (2000). *Design Rules — The Power of Modularity*, Vol. 1. Cambridge, MA: MIT Press.

Baldwin, CY, C Hienerth and E von Hippel (2006). How user innovations become commercial products: A theoretical investigation and case study. *Research Policy*, 35, 1291–1313.

Biemans, W (1991). User and third-party involvement in developing medical equipment innovations. *Technovation*, 11(3), 163–182.

Bladerider International (2007). Bladerider overview. URL: http://www.bladerider.com.au/Press/Valencia/BR0086_bladerider_description.pdf [19 March 2008].

Braun, VR and C Herstatt (2007). Barriers to user-innovation: Moving towards a paradigm of "License to Innovate". *International Journal for Technology Policy and Management*, 7(3), 292–303.

Braun, VR (2007). Barriers to user-innovation & the paradigm of licensing to innovate. Ph.D thesis, Hamburg University of Technology.

DeMonaco, HJ, A Ali and E von Hippel (2005). The major role of clinicians in the discovery of off-label drug therapies. *MIT Sloan*, Working Paper.

Dosi, G, L Marengo and C Pasquali (2006). How much should society fuel the greed of innovators? On the relations between appropriability, opportunities and rates of innovation. *Research Policy*, 35(8), 1110–1121.

Enos, J (1962). *Petroleum Progress and Profits: A History of Process Innovation*. Cambridge, MA: MIT Press.

Faraj, S and M McLure Wasko (2001). The web of knowledge: An investigation on knowledge exchange in networks of practice. URL: http://opensource.mit.edu/papers/Farajwasko.pdf [2 January 2008].

Franke, N and S Shah (2003). How communities support innovative activities: An exploration of assistance and sharing among end-users. *Research Policy*, 32, 157–178.

Franke, N and E von Hippel (2003). Satisfying heterogeneous user needs via innovation toolkits: The case of Apache security software. *Research Policy*, 32, 1199–1215.

Füller, J, G Jawecki and H Mühlbacher (2005). Innovation creation in online basketball communities. Working Paper, Universität Innsbruck.

Harhoff, D, J Henkel and E von Hippel (2003). Profiting from voluntary information spillovers: How users benefit by freely revealing their innovations. *Research Policy*, 32, 1753–1769.

Hars, A and S Ou (2002). Working for free? Motivations for participating in open-source projects. *International Journal of Electronic Commerce*, 6(3), 25–39.

Herstatt, C (1991). Anwender als Quellen für Produktinnovation. Zurich: Rechts- und staatswissenschaftliche Fakultät, Universität Zürich.

Hienerth, C (2006). The commercialization of user innovations: The development of the rodeo kayak industry. *R&D Management*, 36(3), 273–294.

International Moth Class Association of Australia (2008). URL: http://www.moth.asn.au [20 March 2008].

Jeppesen, LB (2005). User toolkits for innovation: Consumers support each other. *Journal of Product Innovation Management*, 22, 347–362.

Lettl, C, C Herstatt and HG Gemünden (2006). Users' contributions to radical innovation: Evidence from four cases in the field of medical equipment technology. *R&D Management*, 36(3), 251–272.

Lüthje, C (2004). Characteristics of innovating users in a consumer goods field: An empirical study of sport-related product consumers. *Technovation*, 24(9), 683–695.

Lüthje, C, C Herstatt and E von Hippel (2005). User-innovators and "local" information: The case of mountain biking. *Research Policy*, 34, 951–965.

Morrison, PD, JH Roberts and DF Midgley (2004). The nature of lead users and measurement of leading edge status. *Research Policy*, 33, 351–362.

Morrison, PD, JH Roberts and E von Hippel (2000). Determinants of user innovation in a local market. *Management Science*, 46(12), 1513–1527.

O'Mahony, S (2003). Guarding the commons: How community managed software projects protect their work. *Research Policy*, 32, 1179–1198.

Ozer, M (1999). A survey of new product evaluation models. *Journal of Product Innovation Management*, 16(2), 77–94.

Raymond, E (1999). The cathedral and the bazaar. *Knowledge, Technology, & Policy*, 12(3), 23–49.

Riggs, W and E von Hippel (1994). Incentives to innovate and the sources of innovation: The case of scientific instruments. *Research Policy*, 23, 459–469.

Rogers, E (2003). *Diffusion of Innovations*. New York: Free Press.

Schumpeter, JA (1942). *Capitalism, Socialism and Democracy*, London: Unwin.

Shah, S (2000). Sources and patterns of innovation in a consumer products field: Innovations in sporting equipment, *Sloan Working Paper* 4105.

Shah, S (2004). From innovation to firm and industry formation in the windsurfing, skateboarding and snowboarding industries. University of Illinois Working Paper '05-0107.

Shaw, B (1985). The role of the interaction between the user and the manufacturer in medical equipment innovation. *R&D Management*, 15(4), 283–292.

Smith, A (1776). An inquiry into the nature and causes of the wealth of nations. Petersfield: Harriman House (2007).

Takahashi, N (2000). The emergence of generalized exchange. *American Journal of Sociology*, 105(4), 1105–1134.

Tietz, R, C Herstatt, P Morrison and C Lüthje (2005). The process of user innovation: A case study in a consumer goods setting. *International Journal of Product Development*, 2(4), 321–338.

Urban, GL and E von Hippel (1988). Lead user analyses for the development of new industrial products. *Management Science*, 34(5), 569–582.

Utterback, JM and WJ Abernathy (1975). A dynamic model of process and product innovation. *OMEGA — International Journal of Management Science*, 3(6), 639–656.

von Hippel, E (1977). Transferring process, equipment innovations from user-innovators to equipment manufacturing firms. *R&D Management*, 8(1), 13–22.

von Hippel, E (1978). Successful industrial products from customer ideas. *Journal of Marketing*, 42(1), 39–49.
von Hippel, E (1988). *The Sources of Innovation*. New York: Oxford University Press.
von Hippel, E (1994). "Sticky information" and the locus of problem solving: Implications for innovation. *Management Science*, 40(4), 429–439.
von Hippel, E (2002). Horizontal innovation networks — by and for users. MIT Sloan School Working Paper No. 4366-02.
von Hippel, E (2005). *Democratizing Innovation*. Cambridge, MA: MIT Press.
von Krogh, G and E von Hippel (2006). The promise of research on open source software. *Management Science*, 52(7), 975–983.

INTERMEDIARIES, USERS AND SOCIAL LEARNING IN TECHNOLOGICAL INNOVATION

JAMES STEWART
University of Edinburgh, UK

SAMPSA HYYSALO
University of Helsinki, Finland

Introduction

Innovation around the Internet in the last few years has stimulated considerable interest in the role of users and user-communities in innovation processes. These currents have resonated with ideas such as 'open innovation' among networks of innovating firms. As the research community is trying to come to terms with these emerging trends, we argue that it can very usefully learn from earlier work on ICT-innovation, the role of users in innovation and the activities of intermediaries in linking of users into supply-side innovation.

There is a huge range of intermediate actors working in between the developers of technologies and their eventual users who do not fit in to conventionally opposed categories such as 'producer' and 'consumer' or 'developer' and 'user'. These intermediaries include retailers, media companies, telecom platform operators, advertising agencies, market research agencies, distributors and management consultancies.[1] (Bessant and Rush, 1995; Howells, 2006). These actors are key players in what new sociology of markets (Callon, 1998; MacKenzie, 2006) calls an 'economy of qualities' by which the needs and desires of consumers are shaped and products adjusted (Callon *et al.*, 2002). This shaping — say between the early demos of pop-music artist and the song in the stereos of his eventual audience — in many ways constitutes technology through packaging, distributing, assembling, quality assurance and testing and branding. Likewise, the 'consumer' is shaped by

---

[1] And equally venture capitalists, lawyers, trade associations, promotional agencies, export agencies, standards agencies, regulatory agencies and so on.

intermediary actors involved in segmenting, persuading, selling, advising, studying and regulating the consumption and in doing so, creating attachment to consumed items, for instance to a branded juice bottle coming from particular orange grove. Instead of an 'invisible hand', it is these very tangible networks that are recently raised to the fore as being able to shape, respond to and maintain seemingly abstract characteristics such as styles and tastes (Hennion, 1989; Callon *et al.*, 2002).

However, innovation studies are particularly interested in new types of products and novel uses, and not just rather stabilised markets and products such as orange juice and pop-music. The complexity of intermediation in innovation networks tends to be underestimated by both practitioners and socio-economic research alike (Stewart, 2007). A major problem with knitting together these networks is that the players involved often have very little previous contact with and understanding of the situations of other players in a nascent market. This can be especially acute between technology developers and the eventual customers and users of the systems. In such uncertain markets, intermediaries play a crucial role, but the mechanisms and contexts of their mediation can be fragile and difficult to predict (Russell and Williams, 2002; Hyysalo, 2004; Williams *et al.*, 2005).

In this paper, we focus on these types of settings and actors we term *innovation intermediaries*: actors who create spaces and opportunities for appropriation and generation of *emerging* technical or cultural products by others who might be described as developers and users. The aim of this paper is to give a framework for addressing the question on how new markets and networks are formed between suppliers and users in the development and implementation of new technologies, at times when existing intermediary institutions may not have relevant expertise or interest. We do this by reviewing the literature on intermediaries and an illustrative review of some of our case studies.

We proceed by first giving some background on how innovation intermediaries are currently understood, and then outline our social learning in technological innovation (SLTI) approach that highlights the importance of use and innovation by users in multi-cycle, multi-level innovation processes. We then map some of the typical intermediaries operating between supply and use, discuss the roles intermediaries play in the shaping of new technology and how they are in turn shaped. In the final sections of the paper, we use SLTI insights on the crucial differences that are in user-involvement in a range of innovation contexts and discuss what this entails for the role and identification of intermediaries. We emphasise how the emergence of appropriate intermediaries is in itself a key part of the overall sociotechnical innovation process and user involvement. We equally stress how some of the crucial intermediaries tend to occupy a fragile position that requires nurturing and protection.

Innovation Intermediaries

Innovation intermediaries can be identified by their engagement in activities in which they gather, develop, control and disseminate knowledge, collect and disseminate financial, technical and institutional resources such as the support of users and sponsors, and attempt to regulate uses, developments, participation and the actions of others in the innovation networks. The extent to which they do this depends on their own access to resources and their connections in the 'constellation' of actors associated with a particular project or emerging market. These intermediaries can be organisations, or individuals grounded in an institutional, technical and often physical context that facilitates their activities. They attempt to configure the users, the context, the technology and the 'content', *but they do not, and cannot define and control use or the technology.* Two crucial features of the environment that innovation intermediaries engage with are: (1) the unpredictability of technological change, market organisation and user uptake and (2) an absence of existing linkages between potential users and suppliers that need to be created in order or innovation to occur and be sustained.

Research on intermediary organisations in innovation such as consultants and other technology brokers has been developed since the early 1990s (Bessant and Rush, 1995; Hargadon and Sutton, 1997).[2] During this period, models of innovation were rapidly changing from fairly linear ones to ones emphasising uncertainty, shifts and the complex interactions between multiple actors that comprised the iterative series of developments jointly resulting in innovation (Freeman, 1979; Kline and Rosenberg, 1986; Williams and Edge, 1996; Van de Ven et al., 1999). The changes in the models were spurred by the growing body of findings about user initiated innovation (e.g., Pavitt, 1984; von Hippel, 1988) and the continued innovation in use (e.g., Rosenberg, 1982; Gardiner and Rothwell, 1985). Analyses of the then relatively new and rapidly evolving fields of robotics and computerised manufacturing technology showed that the talk of diffusion of generic systems matched poorly with the extensive adaptations and further developments done by adopter organisations (Fleck, 1988; 1994; Bessant and Rush, 1995). In short, when

---

[2] We chose to use the concept of intermediary as it has become an established term in the literatures on innovation and organisational studies (e.g., Bessant and Rush, 1995; Howells, 2006) and deployed also in science and technology studies (e.g., Hennion, 1989). Throughout our discussion, we emphasize how these intermediaries change intent, meaning and form of technology through their acts of mediating it between various actors. In this capacity, they are mediators in a sense described by Latour (2005), who uses the term intermediary to denote actors who do not change knowledge or object that simply flows through it. This discrepancy in usage between certain actor network literature and most other literature is unfortunate but there is not much that can be done about it anymore.

the producer company lost its position as the privileged source of innovation, it became urgent to understand how the knowledge from a range of actors flowed into the innovation process.

Consequently, the activities and roles of various intermediary organisations such as consultancies, state development agencies etc., have received attention in various literatures, including innovation management (e.g., Hargadon and Sutton, 1997; McEvily and Zaheer, 1999), literature on innovation systems (e.g., Stankiewicz, 1995) and science and technology studies (Procter and Williams, 1996; Van der Meulen and Rip, 1998; Callon *et al.*, 2002). This interest has also been spurred by the development and growth of knowledge intensive business services (KIBS) that play important intermediary roles (Howells, 2006). Diffusion studies have stressed the importance of change agents and opinion leaders in the diffusion of innovations (Attewell, 1992; Rogers, 2003), and emphasise the work these actors do in tailoring and adjusting the innovation to different audiences and promoting re-inventions that make it more appealing for each particular audience (Rogers, 2003). From a more generic perspective, social network studies have also begun to show the importance of network "bridgers" in not only transferring knowledge across structural holes in networks, but as an important source of innovation themselves (Burt, 2004).

However, to our knowledge there are few studies and frameworks that address in detail the whole range of intermediaries and intermediation that transform technologies, uses and qualities in both use and development domains, and explicate the bridges and gaps that exist in different *ecologies of intermediation* between design and uses. National innovation systems literature aims at this (Lundvall and Johansson, 1994; Stankiewitz, 1995), but only at a fairly coarse granularity, and without analysis of the detailed processes of the learning economy and the substance of this learning (Miettinen, 2002; Stewart and Williams, 2005).

We thus turn to the SLTI framework that allows us to explore in more detailed fashion the dynamics through which intermediaries affect ICT innovation in different socio-economic contexts and constellations of actors with different capabilities, commitments, cultures and contexts (Williams *et al.*, 2005). The framework directly addresses situations of high uncertainties and information asymmetries involved in 'choosing' or 'creating' the right intermediaries for inventive technologies or new groups of users. There is simply more at stake than enabling or preventing the technology from diffusing from suppliers to users.

Social Learning in Technological Innovation

SLTI is a relatively recent approach developed out of the tradition of "social shaping of technology" approach (Williams and Edge, 1996; MacKenzie and Wajcman,

1999) by combining it with insights from other research fields.[3] The development of new technology is characterised as an uncertain process, characterised by complexity, contingency and choice (Williams and Edge, 1996). It places particular design episodes within multiple, overlapping cycles of development and implementation (Rip *et al.*, 1995), focusing on understanding the coupling between technological and social change, and the difficult and contested processes of learning that are integral to innovation.

This analytical framework is socio-technical, and accounts for both technological innovation, and the processes of negotiation and interaction that occur between diverse networks of players attempting to make technologies work — "fitting them into the pre-existing heterogenous network of machines, systems, routines and culture" (Sørensen, 1996).

Many contemporary technologies, particularly ICTs, are not discrete self-contained systems, but "configurations", consisting of layers of components, systems, applications and content, bringing with them partially formed routines, concepts of users and uses, rules for use and other non-technical features. Fitting the existing and the new together involves often long and drawn out relationship building and stop-start processes of institutional learning and forgetting that occur across a constantly changing network of actors.

To understand these processes, the SLTI approach draws together a range of generic mechanisms in which we see learning-though-innovating occurring: learning-by-doing and using in the often trial-and-error processes of appropriating new technologies (Arrow, 1962; Rosenberg, 1982); learning-by-interacting (Lundvall, 1988; Cornish, 1997), as new technologies bring diverse networks of players together; and learning-by-regulating (Sørensen, 1996), as particular players attempt to assert their power though non-technical rules and regulations shaping the 'rules of the game' from everyday use to state policy. These processes — and more detailed learning dynamics within them — not only shape technology, but can have a dramatic effect on the structure of the innovating network, the constitution of the organisations involved, and the identities of the actors (Hasu, 2001; Russell and Williams, 2002; Hyysalo, 2006). Many of these actors and institutions are end and intermediate users, and other societal actors such as governmental and non-commercial institutions. SLTI stresses the importance of giving more detailed accounts of how these actors play key roles in innovation in the long term.

---

[3] SLTI draws on a range of research fields: cultural studies of artefacts and marketing, engaging with the consumption of goods and services; innovation studies stressing non-linear and heterogeneous innovation processes; and work on organisational learning and the reflexive activities of players in the innovation process.

The SLTI approach is thus not a narrowly cognitive, social or modelling process, and the term 'social learning' is used in a very different way to its usage in education and social psychology such as that of Bandura (1977). In the socio-technical usage, social learning denotes the *reflexive* yet often negotiated, complex and 'political' processes in transforming environment, instrumentation and work, that reach beyond single groups of actors. This usage also differs from more generic conceptions of social learning in evolutionary economics (e.g., Wolfe and Gertler, 2002), where learning tends to be taken as an explanatory term for growth in learning economy (Lundvall and Johansson, 1994) without its micro-scale mechanisms and social dynamics being examined (von Hippel and Tyre, 1995; Miettinen, 2002).

Central to the innovation processes identified in SLTI are the creation and evolution of *representations of users and uses*, and their *translation* into technological designs and social actions. These processes are fundamental in shaping design and relationships in the constellation of actors. Far from being solely an up-front 'user needs and requirements capture' process conducted by designers, creation of these representations continues throughout multiple generations of product development. The 'user' is a complex idea: on the one hand, it is a category used by engineers and developers to refer to those who may eventually use their systems, and on the other hand, it can refer to a range of other individuals and institutions, imagined and real, some of which begin to develop various kinds of engagement with a technology over time.

There are many different 'users': intermediate users, end users and proxy users, all of whom can play more or less active roles in articulating their own requirements, and in the creative process. The ability and willingness of developers to engage with these users, and for users to engage creatively with developers is thus central to success, but often extraordinarily difficult. We use the term intermediate users to refer to a particular sort of *intermediary organisations* that adopts a technology for their customers or employees (but generally involving a relatively few individuals within that organisation). Examples are mobile phone operators, banks, retailers who sell a service based on a technological system to end users, and any firm adopting a system to be used by their employees. A subset of these are innovative 'content developers' or content service providers, for example, a service provider such as a broadcaster or publisher offers both a delivery platform and added content for end users. These organisations can be seen as supply-side or demand side within an evolving market according to the particular case and particular point in the innovation and implementation process.

Studies of innovation clearly show that there is a whole range of innovation activities that take part on the user side, particular in early moments of technological change and adoption. Even with comparatively stable technologies and use situations, there can still be innovation by users. Thus, the SLTI approach highlights that many activities and situations that are not conventionally included in the definition

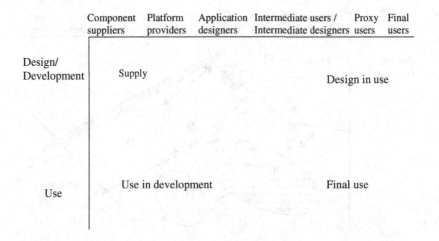

Fig. 1. To account for the users in innovation we need to clarify how they are positioned between primary supply organisations and primary user organisations of a given technology as well as in regards to developing and using (adapted from Williams et al., 2005).

of innovation, are in fact important moments in innovation cycles. This conceptualisation of the role of users in the innovation processes involves moving the focus of innovation studies from the supply-side towards the demand-side. In particular, we need to examine how constellations of players, intermediaries and intermediate and final users, constitute the demand side in the early stages of innovation. This includes examining how they develop uses for technologies and their role in feeding back their experience, practice and innovation to the supply side over multiple long-term innovation cycles.

Mapping Intermediaries Between Supply and Use

Howells (2006) describes the range of different players that mediate various aspects of innovation. Bessant and Rush (1995) go further by elaborating how the range of consultants between suppliers and users of automated manufacturing technology (AMT) each had somewhat different competencies, motives, pricing, clientele and the niche that they occupied in this innovation context. None covered the range and depth of functions that met the needs in emerging areas of innovation. In a similar fashion, Hargadon and Sutton (1997) show how the knowledge-brokering role and industry position of design consultancy IDEO changed as it accumulated more know-how about different industries.

Before moving deeper into the intricacies in the positioning of various intermediaries, let us tentatively sketch some typical intermediaries and their position between supply and use.

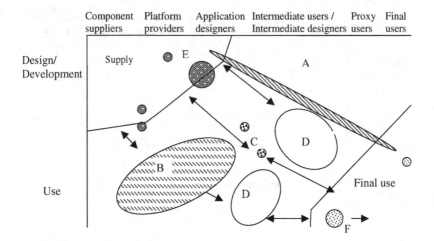

Fig. 2. The niches of some common types of intermediaries between supply and final use.

Figure 2 illustrates the differences in profiles and in consequent mediating capacity of intermediaries. Some intermediaries, such as maintenance organisations and technicians, have long "reach" between supply and use but may be fairly limited in terms of content they cover. (in the figure marked by A). As we found in our study of a health monitoring technology for the elderly (Hyysalo, 2004), technicians can have direct contacts with developers, intermediate users, other supplier representatives as well as end users, but their mandate, interest and expertise remains limited to reporting and fixing technical shortcomings and technical assembly, thus leaving aside questions of marketing, instructing, learning, new uses etc. The "width" of their mediation thus tends to remain narrow, unless their task description gets expanded to include other tasks such as user-training, as happened at one point in the case studied (Hyysalo, 2006).

An example of a broad width but substantially shorter reach into both use and supply side would be retailers (in the figure marked by D) and Telecoms operators (in the figure marked by B). Such actors exercise competence and power over multiple technologies and several key aspects of technology such as pricing, distribution channels, marketing, branding, feedback from other intermediaries and end users etc. Yet another sort of intermediaries are market research and usability consultants (in the figure marked by C), that accumulate, refine and transfer (second order) information both about products as well as of their usages. The most studied type of intermediaries are various supply-side industry consultants (in the figure marked by E), who may play central roles in augmenting innovation at supply end and passing inter-industry insight. These include engineering and business consultancies, public research agencies, industry contractors, accreditation agencies etc. (e.g., Bessant and Rush, 1995; Van der Maulen and Rip, 1998; Howells, 2006).

Intermediaries at the supply-side business to business environment tend to be more numerous, visible and formal than those close to the end-users of consumer goods. Amongst the use-side intermediaries (marked by F in the figure) those involved in buying and the paying for new technology are relatively more visible than those that help people use, fix, maintain and update their technologies. The latter are often less formal and may perform their work as peer favours or sidejobs to their formal work. As a consequence, it is these intermediaries that are particularly systematically neglected or underestimated. It is indicative that discussions of such peoples as "local experts" (Stewart, 2007), "technology mediators" (Okamura *et al.*, 1994) or "tailors" (Trigg and Bodger, 1994) remain absent from technology management volumes that abound with literature on product champions, business angels etc., at the supply end.[4]

The asymmetric distribution of knowledge amongst actors has the result that people and organisations that hold intermediary positions tend to accumulate increasing amounts of the kind of knowledge that flows in from their various clients and projects, whereas other actors do not. The net result is that less central actors (such as new supplier entrants, end users) face difficulty in assessing the landscape, position of different actors within it as well as the means at the disposal of those actors to hinder or enable the prospering of new technology. In fact, such structural holes and knowledge asymmetries are crucial in the existence of the very niche of many actors, and we return to discuss this theme in more depth below (Burt, 2004).

What our diagrams let us articulate is that there is a range of "short and fat" intermediaries whose high breath but short reach allows linking fairly homogeneous actors that are already quite close to each other in the supply network, or in user communities. There are also "long and thin" intermediaries that have high reach in linking users and suppliers through specialising in a particular service, but consequently only limited breath of mediation. However, it appears to be quite hard to be a "long and fat intermediary" with both broad scope and long reach. Certain large-scale retail organisations (such as Wall-Mart) show that this is possible in mediating incremental innovation, but in more uncertain and evolving markets examples known to us are few and debatable. The reason may be simple. Intermediaries are boundary spanners conveying delicate, sticky knowledge, conducting negotiations that need trust, setting rules of use that require legitimacy and configuring technologies and so on all of which requires specialist knowledge. It may be

---

[4] Although they have attracted more attention in marketing around opinion leaders, word of mouth and viral marketing (Stewart, 2007).

hard to scale up these capabilities up in both breadth and range given that real-life intermediation is a fluid and delicate phenomenon.[5]

In general, what the early innovation studies accounts of intermediaries largely failed to address was that established intermediaries can also be roadblocks, and expensive and intransigent gatekeepers, with services, repertoires of knowledge and activities, that can *fail* the innovation process in a range of ways.[6] Let us illustrate this with an example in the area of video games for girls and women. The established industry of games publishers and events, magazines aimed at existing market for these products is almost wholly devoted to promoting particular range of game genres to a young male market. For a firm which identified a market for "girl games" and is able to engage with potential users in the design of attractive products, these intermediaries are not a resource but a hindrance, and necessitate re-casting the products and making new connections to non-traditional intermediaries — such as general retailers, museums and TV broadcasters. And also in real life such a shift proved necessary for success (Stewart, 2004).

What Do Intermediaries Do in Social Learning?

Perhaps the clearest way to approach the range of activities in which intermediaries are involved is to first look at some taxonomies that exist in the literature (Bessant and Rush, 1995; Hargadon and Sutton, 1997; Howells, 2006). Howells suggested 10 functions for innovation intermediaries, even though he admits that individual intermediaries seldom play separate functional roles, but contribute and develop a range of different activities important in innovation. In a similar vein, Bessant and Rush (1995) list six bridging activities through which consultants bridge suppliers and their customers. These activities develop not just through working on one-off projects, but by developing long-term capabilities of the individual firms, and of the market as a whole. These consultants tend not to work only on a triad basis but are generally involved in several relationships.

These typologies of functions and activities of intermediaries approximate the generic terrain of intermediaries. However, as Bessant and Rush point out, there is work to be done in charting the roles that intermediaries play within these functions

---

[5]Thanks to Robin Williams for this formulation.

[6]In contrast, other work from evolutionary economics and related perspectives has discussed the way in which innovation systems may operate as a selection environment, weeding out challengers that do not fit with the established technology regime (e.g., Nelson and Winter, 1977; Dosi, 1982; Rip and Kemp, 1998).

Table 1. Functions and activities of intermediaries.

Intermediary functions (Howells, 2006)	Bridging activities (Bessant and Rush, 1995)
1. Foresight and diagnostics	1. Articulation of needs, selection of options
2. Scanning and information processing	2. Identification of needs, selection training
3. Knowledge processing and (re)combination	3. Creation of business cases
4. Gatekeeping and brokering	4. Communications, development
5. Testing and validation	5. Education, links to external info
6. Accreditation	6. Project management, managing external resources, organisational development
7. Validation and regulation	
8. Protecting the results	
9. Commercialisation	
10. Evaluation of outcomes	

and activities.[7] All these intermediary roles involve knowledge creation, translation and dissemination. They are all also about making a connection between memory/experience and future visions, and instantiating these two in current actions of the people whose actions are mediated by them. When we try to differentiate fundamentally different facets in the actions of intermediaries, three distinct roles in social learning become salient: facilitating, configuring and brokering. These more generic roles are better applicable to the range of intermediaries in social learning processes between supply and use. We anchor our discussion at the intermediary roles that cybercafés played in the mid-1990s when the Internet was relatively new — characteristically user-end intermediation (Stewart, 2000).[8]

Facilitating

Facilitating can be described as providing opportunities to others, by educating, gathering and distributing resources, influencing regulations and setting local rules. Facilitation involves "creating spaces" of various types: social (communities, networks), knowledge (skills and know-how resources), cultural (positive images), physical (a place or equipment), economic (providing funds), and regulatory

---

[7] Bessant and Rush (1995) identify four generic roles, those of transfer of knowledge, sharing knowledge across user community, acting as brokering to a range of suppliers and diagnostic/innovation role in trying to identify what end users actually want.

[8] Cybercafe and Internet centre innovators took computers and the Internet out of offices and homes, and put them into a new context, introducing them to new users and providing a new setting for existing users. What was considered at the time a fleeting and unimportant configuration of the technology involved considerable local service innovation, and has since become an extremely popular and successful service model.

(creating rules to guide activities and reduce uncertainty). In the case of cybercafes, the role of cybercafés and their managers as facilitators is clear (Stewart, 2004). The cafe is a convenient and open, friendly physical space, conveniently located, with an informal atmosphere, in which the managers had developed based on their initial concept of users and uses. They provide the computers and software, and the training and advice that is needed to use it. The expertise and knowledge that they supply to the users is as important as the actual technology. They take the headache out of computer use, and create a flexible environment where people can work, play or learn at their own discretion. Training and informal support and the creation of an atmosphere that encourages interchange between users are important facilitation devices. Another important facilitation role is running trials that generate new interactions between users and suppliers, and importantly, make the activities and results visible in wider to outside actors. Of course, the cybercafe is a literal space, but we have seen a huge growth in industry-user fora, user and industry networking groups, conferences and seminar series, various government and private funds for experimentation and interaction, and creation of regulatory spaces providing temporary protection from regulations and rules usually applied in a particular environment.

Configuring

The creation of the space that facilitates appropriation by others and influencing the perceptions and goals of sponsors and users involves active processes of *configuration*. This includes configuring technology, often in a minor way; creating and configuring content; setting rules and regulations on use and usage, prioritising uses, the goals and form of projects, and the goals and expectations of other members of a network. Configuration is not only technical but also symbolic; intermediaries provide an interpretation of the product, the meanings that people give to a technology, but they also listen to users, sponsors and suppliers and attempt to modify the project to reflect their interpretations.

The managers and owners of the cybercafes in the cases did not invent the cybercafe — computers in cafes were not a new idea at the time of this study. However, they had to make decisions about what a cybercafe was, what was relevant to them and their business, and to their customers. This business model led to the configuration of the space, rules of use, configuration of computers, and policy on what users to encourage or discourage. This included the appropriate types of uses: games, the Internet, office service etc., for their café and clientele (Laegran and Stewart, 2003). However, this was not necessarily a one-off configuration: it changed rapidly as customers introduced their own ideas of what a cybercafe should be, bring practices in from outside, and evolving them from within. Some cafe managers really took on board the need for constant re-configuration and experimentation while others evolved a much more stable model, with little space for user-led change. The cafes also attempted to configure their customers' usage of the cafe through information, training, and informal learning, and introduce new users, for example by running classes for women or older people. By encouraging new uses and new users, they are, of course, encouraging people to spend more time in the cafe, but also making sure that they can appeal to more people, and help customers diversify their use. Of course, in order to do these types of activities intermediaries such as cybercafés and their managers have to gain legitimacy, but this can be self-fulfilling if their configuration activities are successful.

Brokering

The third activity of intermediaries in social learning processes is brokering. For example, intermediaries act to raise support for the appropriation process from sponsors and suppliers. They set themselves up to represent appropriating individuals and institutions, and negotiate on their behalf. Intermediaries need to broker entry of new sponsors or suppliers into their project in order to defend the space they have helped to create, and make sure that they increase their access to resources and knowledge and can maintain influence over rules and practices. Some of the brokering activities can be around the features and functionalities of new technologies, directly communicating needs and requirements of users and the possibilities and conditions of change of the suppliers.

In the cybercafes case, the manager of one community cybercafe had a strong role as a broker. The cafe came about as a result of his relationship with the funding council, the local community groups, sponsoring companies and local and national politicians. The project was rather outside the mainstream community project, and certainly not a business for which he could get a bank loan for, so his negotiation with sponsors, as suppliers of equipment, money, prestige was the only way to make it happen.

Brokering is certainly one of the most direct ways that intermediaries can bring users and suppliers together, but as this example shows it is equally important in bringing other important actors into the local innovation network, and maintaining their commitment and interest, while at the same time communicating the importance of the particular innovative process to their interests. One of the key balancing acts they have to manage is maintaining the openness of their facilitation activities in the face of the brokering activities: after all brokering is rather a heterogeneous bridging position than representing a particular interest.

In the case of the cybercafes, one set of intermediaries — the managers — were involved in all three processes, and similar functions, such as training, all played a role in them. While many intermediaries may focus on type of activity, particularly in stable environments, the dynamic and unpredictable nature of innovation can lead them to conduct all three. Intermediaries that are likely to be most successful can enter into and balance different activities without constraining the innovative activities of their clients, be they adopters or suppliers. The affinities between the intermediation by cybercafés and their managers in early 1990s and present day bounded socio-technical experiments such as living labs are noteworthy.

Pre-Domestication, Power Games and Fragile Intermediaries

There is a range of ways which intermediaries influence *the evolving shape of technology*. First, when "local experts" and "tailors" do work of brokering, facilitating and configuring, they prefer certain options and suppress others in their effort to cater a system that is practically useful and usable for the particular user or organisation as was evident in both cybercafé and health-monitoring cases. In turn, this work tends to rely on other intermediaries, such as the products of service and technology suppliers, specialist magazines, web-pages etc., and eventually translates into supplier offerings. By doing so, intermediaries are engaged in *pre-domestication* — influencing what would be an appropriate target for the ongoing development of technology, what could be the appropriate goals and motives for using it, and making technology appropriable in their practice. This ordering of a potential development arena can grant them a position for informal learning by regulating (cf. Sørensen, 1996).

While flagging the importance of enriching and shaping of the technological offer, power and influence issues need to be recognised. Enrolling other players means selling the technology to them. Distributors, operators etc., have their own perception of user needs, and have different interests and incentives than the supplier or end users in promoting some products and not others, in pricing, branding and in aligning products. The technology thus gets framed for intermediary audiences in addition to its assumed final consumers. The product, especially widely distributed content products like games or books, has to be first sold to intermediaries such as a distributor to ever reach the final consumer. Making technological products or services appeal to intermediary audiences affects it beyond sales arguments or other "wrapping" and tends to cut into features, functionalities and look of the product. For instance, in games development small companies view the publishers as their primary customers, and anticipate their selection processes alongside (or even instead of) that of end-gamers. The assumed norms and extrapolations over previous behaviour of key institutionalised intermediaries thus channel design already before it ever reaches them directly (Kalhama, 2003; Eskelinen, 2005).

Intermediaries often work hard to make themselves obligatory points of passage (Latour, 1987). Commercial firms may do this as the basis of their business model (Burt, 2004), but other intermediaries, such as trade associations or user groups may also do this in an attempt to established a strong bargaining position for their members. In the domain of IT for example, the Gartner Group has established a key role in setting up expectations for the future of a software application sector, becoming guardians of community knowledge and thus a key mediator in shaping the behaviour of suppliers and users (Pollock and Williams, 2008). However, it is a delicate balance — they must remain accountable to those they service and represent.

The above dynamics get more complex through the uncertainty regarding markets and users' preferences for new technology prior to its actual usage (Hyysalo, 2003; Williams *et al.*, 2005). The need for or effects of different framings of technology are not readily visible at the outset to any of the parties. Images of users and customers become "currency" that is proffered and sold to establish and contest business cases (Nicoll, 2000). Indeed, the ability of intermediaries to cut the cake is dependent on how convincingly they can argue their importance and hence, their vision of the user and the buyer. This is not unlike the way intermediaries offer assembly and maintenance services that convince the users of images of a technology that is too cumbersome or impossible for users to handle themselves.

It is common to use a range of more-or-less publicly available (often grey literature) sources such as newspaper headings and consultancy reports as "external" legitimizing devices for arguing the case for one's own technology and vision about the own and user domain development, as well as doing one's best to influence them. However, because these images circle and contest one another, "real user data" such as that from usability studies tends to be "hard currency" (Nicoll, 2000) in comparison to market studies and other inferred proxies. Various trials, pilots and demonstrations become instrumental for different parties arguing their case and relevance.

By the same token, it needs to be underlined that some potentially very useful intermediaries operate under rather adverse conditions, as their intermediator roles hinge upon corporate policies and reward structures that have a bearing upon what roles people can take on. Research on two formal roles, sales teams and technical support staff is illustrative. Williams and Proctor (1998) found that IT support staff had the closest ongoing relationship with users, and clearest awareness of problems, and knowledge of on-the-ground innovations in use. However, these were also the people most distant from the future product designers within the supply sector. In the same token, the sales people were seen as bridges to developers, but mostly talked with purchase people, and not with end users. Also the incentives for conveying information to product development tend to be lacking: sales teams are rewarded on the basis of the deals they close, not on the potentially helpful R&D information they may gleam from customers. Reward structures that would encourage side bets relevant to social learning may also prove rather difficult to set in place without undermining the effectiveness of sales-based structures.

This tangle gets more tricky when the sales are handled by different organisations: getting such sales staff to sell a conceptually new product is one thing, getting them to do this in a desired manner is another. In the health-monitoring case, the supplier continuously struggled to ensure that the various sales peoples had adequate understanding of their product, targeted preferred customer segments, did not make inadequate sales promises and transferred more needy customers to supplier's sales people who had sufficient expertise. Even more formidable problems persisted in getting them to glean and pass on information about customers (Hyysalo, 2004). In the words of the company president:

> *"Since [our new customer support and maintenance person] started, it has turned out that our retailers, partners, and assemblers haven't really provided us with information about how the device works in actual use. Neither do they always know how the device should function... Here is the one employment that has most effectively paid for itself"* (Interview with the company president 17.9.2001).

This quote also introduces another issue: intermediary roles are often carried by individuals. While established intermediary organisations can institutionalise intermediary actions, in new areas these are likely to be vested in individuals who have expertise and bridging links into other areas of activity, gathered by earlier career moves, or though some other personal contacts and experience. The person talked about in the quote above had held a range of similar roles in maintenance, training and marketing in several other companies in safety-phone business for over 10 years. Our further interviews revealed that over a two decades there had been (just) three people who had remained in the safety phone technology business by circulating between different suppliers in sales, assembly and maintenance and management posts. These three had become the living repositories of the accumulated learning on how to best deploy and hold in operation over 50,000 systems and various new kinds of entrants to this clientele. (Hyysalo et al., 2003).[9]

---

[9] We note anecdotally from years of observing specific domains of innovation that such circulation within limited space is common, as individuals move between firms and projects acquiring increasing

One of the reasons to emphasise *social* and *learning* in technological innovation lies in the need for actors to participate in the ongoing circulation between development and use. Intermediaries are continuously forced to learn about, filter, translate and reflect on information, products and practices of other actors to remain relevant and thus in existence. An important part of this learning is about how to relate and manipulate as well as how to dominate and control other actors around them. But how this ecology is set up can differ dramatically depending on the innovation context in question; this sets the relationships between development, use and user-involvement in innovation.

Innovation Contexts, Social Learning and Intermediary Roles

Attending to the variations in innovation spaces is one of the key features of SLTI framework (Williams *et al.*, 2005). There are remarkable differences in the degree of freedom for innovative actors, particularly users, to try out new things, exercise choice, or act reflexively (Bessant, 1991). At one extreme, users remain relatively "passive" with little choice over adoption. This is the much criticised "linear" innovation model, where users appear as consumers of pre-formed technologies, where their only choice is between use and non-use of a technology. Each member of a supply chain can thus be regarded as an intermediary between the preceding and following player, and end users only have contact with the final player in the chain. Suppliers and end users are separated and user preferences and innovations are signalled at arms length through a market back to suppliers. We can display this graphically (Fig. 3, adapted from Williams *et al.*, 2005), which also clarifies that such market signals may not be very clear, and certainly not to the whole market, and invisible to firms deep in the supply network (Fleck, 1988).

The early years of the health-monitoring project 1993–1999 illustrate well the kinds of intermediaries that tend to be involved in mediating between development and use in linear context. Literature on the users' domain, two market research studies, a design and usability study, pilot-trials and a branding company fed representations of "the user" and appropriate design decisions to the developer company. After the launch of the product, intermediate users such as assembly people from vendors and managers of rest-homes for the elderly became key user-side intermediaries that helped to configure the system and re-configure the associated work practices, facilitated the everyday uses and problems and brokered contacts and information between the everyday usages and suppliers' ongoing development efforts. The

---

expertise and knowledge about the domain specific roles inside a supplier organisation, promotional agency, sales of other company, and procurers in a user-side organisation.

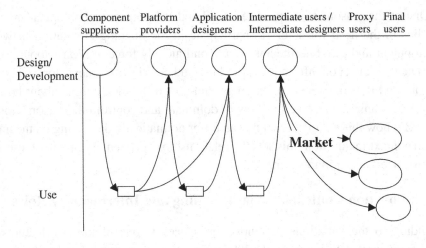

Fig. 3. Pipeline linear development and diffusion (adapted from Williams *et al.*, 2005).

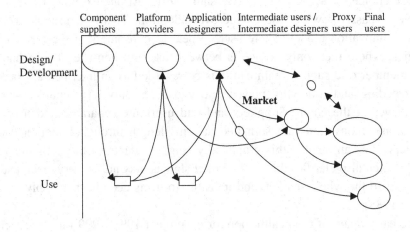

Fig. 4. Intermediaries in linear development and diffusion; the knowledge of the user is mediated significantly by various intermediaries to application developers and user-end intermediaries tend to be involved in configuring and facilitating the usages.

component and platform providers were seemingly bracketed off "behind" the company building the health-monitoring application, but in reality the configuration of the system brought issues around platform and integration with other technologies back in, giving rise to intermediary activities by assemblers and local small vendors (Hyysalo, 2003, 2004) (Fig. 4).

Through a range of case studies done as part of the European *Social learning in Multimedia* project in the late 1990s, Williams *et al.* (2005) distinguish four other modes of user involvement in innovation: the evolutionary "pick and mix" model; user-centred design; the innofusion and appropriation model and a technology experiment.

Evolutionary pick and mix context

The "pick and mix" model is closest to the market model, where intermediate and end users are able to pick from a huge range of available generic technologies, and configure them together. This model is characteristic of the current ICT market, where intense competition, flexible standard platforms such as common operating systems and internet protocols, and open programming interfaces and tools make it relatively easy, and very cheap to configure.

Here, we see the emergence of a range of intermediaries that configure technologies and uses, attempting to bridge the "market gap" from suppliers to user and vice versa. The cybercafés illustrate how well in many cases the user-side intermediaries play crucial roles in configuring, brokering and facilitating users, even establishing wholly novel intermediate-design/intermediate-use locations. The tentative depiction of various kinds of intermediaries given in Fig. 1 matches closest the situation in a pick 'n' mix context (Fig. 6); the ecology of intermediaries is comprised of multiple types of actors and is likely to vary from one product to another so that different sets of intermediaries are involved in mediating the various platforms, components, content and applications.

In this environment, intermediaries can be very sensitive to end users and often have to respond rapidly to their demands and innovations, but equally, end users can be confused and particular reliant on the work of intermediaries.

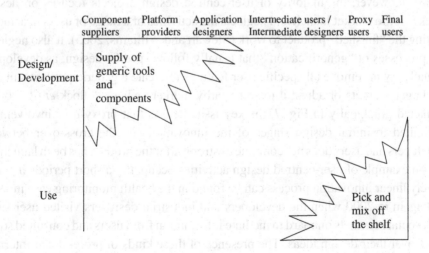

Fig. 5. Pick and mix model where there are large clusters of generic offers at the supply end and the configuration of off-the-shelf components at local user sites. (adapted from Williams *et al.*, 2005).

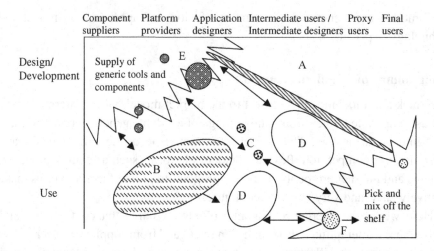

Fig. 6. The niches of some common types of intermediaries illustrated in pick and mix constellation.

User-centred design

There are a number of design and development processes that involve users in more active ways. The first of these is user-centred design processes in which end-users — or more correctly "proxy users" who represent eventual users — are put at the centre of design. Detailed studies of users, along with negotiations with proxy or intermediate users of their "needs and requirements" help those creating new technologies or integrating systems to create products and services that closely match the existing culture and activities of specific users (e.g., Norman and Draper, 1986). However, the majority of user-centred design projects focuses on design work prior to market launch and neglects the activities of a range of users in actually getting the "finished" product to work (Stewart and Williams, 2005). It also neglects the processes of "generification" that usually follow specific design, as developers actually try to remove all specific user features to create a generic product suitable for larger markets or adjust it to suit nearby market niches (Pollock et al., 2007). Depicted graphically in Fig. 7, the key issue here is that proxy-user involvement is limited to initial design stages of the innovation and the cross-over between developers and users does not continue as strong after the product has been launched.

An example of user-centred design activities occupying a short periods in a relatively linear innovation process can be found in the health monitoring case in 1995 and again in 2000 when the developers and industrial designers visited user sites, took social scientists onboard to mediate information from users and consulted some users over their design ideas. The presence of these kinds of professional intermediaries for mediating use to design is typical to user-centred design, as is organising it in projects (e.g., in concept design) both in terms of company practice and more

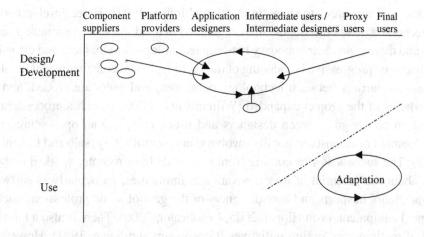

Fig. 7. User-centred design. A more dedicated application is built with the help of proxy users (adapted from Williams *et al.*, 2005).

academic literature (Preese *et al.*, 2002; Dix *et al.*, 2004; Benyon *et al.*, 2005). In the Fig. 7, this is portrayed by the actor below the design project contributing to it.

The Technology Experiment/Evolving Co-Design

The Technology Experiment (Fig. 8) is a mode of collaborative innovation that involves a range of players, such as government agencies, intermediate users, developers and suppliers (Jaeger *et al.*, 2000; Brown *et al.*, 2003), often deliberately constructed into a constituency by certain key players to provide a framework

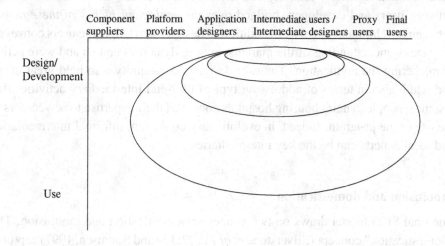

Fig. 8. Technology experiment/evolving co-design project (adapted from Williams *et al.*, 2005).

of ideas and resources to shape innovation (Molina, 1995). Here, development of the technology and building the market go hand in hand: these are continuing activities and there is no clear boundary between technology development and diffusion. There can be progressive broadening of the socio-technical constituency of involved players as barriers between technology developers and users are eroded, and the boundaries of the project expanded (Williams *et al.*, 2005). Such a process can be based on co-design between designers and users, comprise an open-source type development or at least temporally involve users fruitfully (Hyysalo and Lehenkari, 2003). The success stories coming from this mode have recently sparked enthusiasm about the potential of user innovation communities, particularly in software, pro-amateurs in sports and user-designers of the gear of some professions such as surgical equipment. (von Hippel, 2005; Leadbeater, 2006). There is also a long history of participatory design initiatives (Greenbaum and Kyng, 1991). However, a technology experiment can also merely verify the chosen technology model negotiated early on in the process. This partly depends on the degree to which core players are open to innovation by users, and the points at which configurations are locked into place (van Leishout *et al.*, 2001; Hoogma *et al.*, 2002).

In the evolving co-design, intermediaries and intermediation between development and use can differ dramatically from the previously discussed innovation contexts. Let us illustrate with a user-initiated design project to build electronic-health record for diabetes professionals where a group of lead-users enrolled an IT company to join their effort. Their collaboration lasted intensively from 1996 to 2002 and through three development rounds (Hyysalo and Lehenkari, 2003). What is striking for us here was the lack of and irrelevance of intermediaries one encounters in other innovation contexts such as market research and probing, marketing and advertisement agencies, usability consultants etc. In effect, the co-located design between users and developers bridged over *the ecological niches of these actors* between development and use. In addition to design knowledge, the users conveyed very exact understanding of the markets, buying dynamics and so and were active in marketing the application. The user-designer community also held great depth and width also in terms of addressing typical user-end intermediary activities: for instance, people in neighbouring hospitals conveyed their expertise to newcomers to the use of the program. Indeed, in evolutionary co-design, informal intermediaries and local experts can be the key intermediaries.

Innofusion and domestication

The final SLTI model draws on two concepts: domestication and innofusion. The "domestication" concept (Silverstone *et al.*, 1992; Lie and Sørensen, 1997) captures the practical, symbolic and cognitive dimensions in the selection, deployment and adaptation of new technologies. The innofusion concept (Fleck, 1988) highlights the technological innovation done in these processes, emphasising that key innovation moments occur in and are controlled by the user environment. The interactions between networks of users and designers are not continuous or controlled, but are constantly changing, as different sets of actors in the constellation of interested parties are temporarily linked. This innovation context differs from user-centred design in that it is in users' sites, not in prior design where key user involvement occurs and that their innovative inputs can last for years. Innofusion differs from evolutionary co-design in that the relations between producers and users may not be collaborative, purposefully co-ordinated or co-located (Fig. 9).

Examples of innofusion and domestication context can be found in various types of technologies. A recent well-documented case comes from enterprise resource planning systems (ERP) in educational sector (Pollock and Cornford, 2004; Pollock and Williams, 2008). The developer companies made initial customisations to systems built for other sectors. The early customers were involved in further specification of the modules, contents and functionality of the system. Their IT-staff worked further on the system including configuring the package in-site within its myriad of built-in parameters, more extensive customization through re-writing of code, selective appropriation of the package as well as integrating add-on, bolt-on and extension software. Some of these modifications became later incorporated as parts of supplier's generic package, while some became discarded and kept up only locally. In contrast to this drawn-out innofusion interchange between the supplier

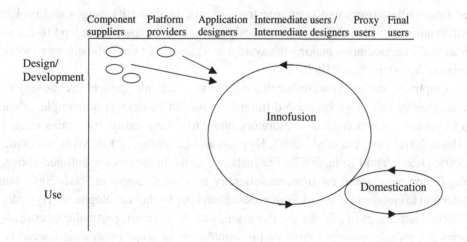

Fig. 9. Innofusion and domestication model (adapted from Williams *et al.*, 2005).

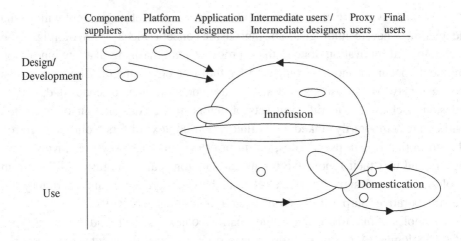

Fig. 10. Intermediaries in Innofusion-domestication context. A range of supplier-dedicated, more independent formal consultants and informal intermediaries tend to be involved.

and early users, many later adopters were effectively confined to more limited domestication in-site having little chance to shape the evolution of the generic system more to their liking (Pollock et al., 2007). Also in this innovation context, a range of intermediaries can be observed (Fig. 10).

There are highly influential industry analysts such as the Gartner Group that actively shape the expectations of the market place, criteria used to select ERP providers, criteria to measure supplier performance etc. (Pollock and Williams, 2008). Closer to the user clients' IT-staff and implementation consultants act as intermediate users that configure the systems, facilitate its usages and broker information between different groups of people. Local expert users tend to act as user-end intermediaries making the system work, not least through helping to work around the often clunky ERP.

Empirical studies show further that over time a particular project, technology or constituency can move between different innovation contexts, for example, when a technology moves from an exploratory phase to generification for a mass market (Hasu, 2001; Hoogma et al., 2002; Hyysalo and Lehenkari, 2003; Williams et al., 2005; Osterloh and Rota, 2007). For instance, in the diabetes case outlined above, the IT-company withdrew from evolutionary user-involvement in 2001–2002, but later had to engage in innofusion-domestication type further development (Hyysalo, 2006; Miettinen et al., 2008). As these kinds of shifts occur, particular intermediaries can switch between formal and informal roles or some intermediaries can be sidelined in favour of others.

Conclusion: Recognising Key Intermediaries and Nurturing Fragile Ones

Within the context of the user-designer relationship analysis in the design literature, and the supply-side – demand-side approach in innovation studies, we highlight the importance of new intermediaries. These actors emerge to bridge gaps in newly forming networks, facilitate contacts and experimentation by passing knowledge, they act as a store for collective memory, and shape technologies, visions, knowledge and relationships. These activities are central to the social learning that occurs in innovation: the processes of creating new relationships and knowledge that accompanies the creation of new technologies.

The SLTI approach allows bridging research on intermediaries at the appropriation end with the more plentiful literature on supply end intermediaries in innovation and organisation studies. Applying this framework to the study of intermediaries underscores five issues that cut across extant literature as well as our case studies over the years:

(1) *There is an ecology of intermediaries in and between supply and use.* The highly visible supply-side intermediaries such as foresight agencies and management consultants, and the easily identifiable middle-ground agencies such as marketing research and usability consultants tend to overshadow the often more informal yet just as crucial intermediaries at the user-end of the supply-use relation. Intermediate users, local experts and "tailors" facilitate, configure and broker systems, usages and knowledge about systems and their deployments, helping users to domesticate them and suppliers to respond to actual, realised uses.

(2) *Pivotal importance of uncertainty and learning.* When it comes to innovative new products no established line-up, or a smoothly functioning chain of intermediaries can be assumed, in contrast to stable products such as orange juice or pop-music. One cannot assume that extant intermediaries will be the relevant ones for successful innovation. Gradual adjustment and learning about the appropriate form of the technology, actors to be involved, expectations concerning its use practices, complementary services, delivery models and other support tends to be more important than the effectiveness in carrying out these activities that established intermediaries may have.

(3) *Important differences are due to innovation contexts.* The knowledge and actions needed to bridge supply and use depends greatly upon the nature of the technology in question, but also on the innovation context where the technology develops. Sustained co-design between developers and users can render obsolete the need for in-between niche players such as market researchers and usability

specialists that are more likely to hold pivotal importance in the success of innovations developed within linear context.

(4) *Identifying and nurturing best suited intermediaries is the challenge.* Charting the intermediaries related to a particular innovative project is a recognised part of what technology managers do. Where they lack analytical clarity is in charting the ecology of intermediaries to gain a sense of how each key aspect of a given technology and key user segment — the market — becomes mediated. Identifying and finding means to sustain people who perform important mediating roles — inside or outside their formal organisation — tends thus fall by the wayside and, according to our findings, if the benefits or appeal of intermediating some line of technology or in some organisation wane, these people are quite likely to shift location.

(5) *User involvement in innovation has a wide range as does its relationship to and dependency on intermediaries.* Ranging between everyday domestication and massive evolutionary opensource development projects, user involvement in innovation has myriad of forms and outcomes. Some of the practices that get discussed in the literature as user innovation may be better understood as user-end intermediary activities in performing the key intermediary roles we identify: configuring, facilitating and brokering new technology. Moreover, in some innovation contexts hardly anything about users' innovative solutions and practices in using technology is conveyed to technology supply without intermediaries. Yet, in others, direct user-involvement can bridge over the niches typically occupied by for instance market research and usability consultants.

These cross-cutting themes are highly relevant for the presently emerging methods such as Living Labs and various experiments in co-creation as these create new forms of user-involvement and intermediary positions made possible with the current ICT infrastructures and standard platforms. Here too, existing intermediaries cannot always provide the necessary links to engage potential users in innovation, and new intermediaries need to be found, nurtured or created. One of the challenges remains to avoid making the mistakes of configuring spaces that are dominated by technological considerations and do not allow space for the participation of existing and new intermediaries. There is still considerable work to be done in understanding how intermediaries can be managed at a policy or corporate level, and how they can be prevented from failing.

References

Arrow, K (1962). The economic implications of learning by doing. *Review of Economic Studies*, 29, 155–173.

Attewell, P (1992). Technology diffusion and organizational learning: the case of business computing. *Organization Science*, 3(1), 1–19.

Bandura, A (1977). *Social Learning Theory.* Englewood Cliffs, N.J: Prentice-Hall.
Benyon, D, P Turner and S Turner (2005). *Designing Interactive Systems — People, Activities, Contexts, Technologies.* London, UK: Addison-Wesley.
Bessant, J (1991). *Managing Advanced Manufacturing Technology: The Challenge of the Fifth Wave.* Oxford: Blackwell.
Bessant, J and H Rush (1995). Building bridges for innovation: the role of consultants in technology transfer. *Research Policy,* 24, 97–114.
Brown, HS, P Vergragt, K Green and L Berchicci (2003). Learning for sustainability transition through bounded socio-technical experiments in personal mobility. *Technology Analysis & Strategic Management,* 15(3), 291–315.
Burt, R (2004). Structural holes and good ideas. *American Journal of Sociology,* 110, 349–399.
Callon, M (1998). *The Laws of the Markets.* London: Blackwell Publishers.
Callon, M, C Méadel and V Rabeharisoa (2002). The economy of qualities. *Economy and Society,* 31(2), 194–217.
Cornish, S (1997). Product innovation and the spatial dynamics of market intelligence: does proximity to markets matter? *Economic Geography,* 73(2), 143–165 (April 1997).
Dix, A, J Finlay, GD Abowd and R Beale (2004). *Human-Computer Interaction.* 3rd Ed. Harlow, UK: Pearson Prentice Hall.
Dosi, G (1982). Technological paradigms and technological trajectories — a suggested interpretation of the determinants and directions of technical change. *Research Policy,* 11, 147–162.
Eskelinen, M (2005). *Pelit ja pelitutkimus luovassa taloudessa (Games and Game Research in Creative Economy).* Helsinki: Sitra.
Fleck, J (1994). Learning by trying: the implementation of configurational technology. *Research Policy,* 23, 637–652.
Fleck, J (1988). Innofusion or diffusation? The nature of technological development in robotics, *Edinburgh PICT Working Paper* No. 7, Edinburgh University.
Freeman, C (1979). The determinants of innovation — market demand, technology, and the response to social problems. *Futures,* (June), 206–215.
Gardiner, P and R Rothwell (1985). Tough customers: good designs. *Design Studies,* 6(1), 7–17.
Gertler, MS and DA Wolfe (2002). *Innovation and Social Learning: Institutional Adaptation in an Era of Technological Change.* New York, NY: Palgrave MacMillan.
Greenbaum, J and M Kyng (eds.) (1991). *Design at Work: Cooperative Design of Computer Systems.* Hillsdale, NJ: Lawrence Erlbaum Associates.
Hargadon, A and R Sutton (1997). Technology brokering and innovation in a product development firm. *Administrative Science Quarterly,* 42, 716–749.
Hasu, M (2001). *Critical Transition from Developers to Users.* Academic Dissertation. Helsinki: University of Helsinki, Department of Education.
Hennion, A (1989). An intermediary between production and consumption: the producer of popular music. *Science, Technology, & Human Values,* 14(4), 400–424.

Hoogma, R, R Kemp, J Schot and B Truffer (2002). *Experimenting for Sustainable Transport — The Approach of Strategic Niche Management*, Vol. 10. London: Spon Press.

Howells, J (2006). Intermediation and the role of intermediaries in innovation. *Research Policy*, 35(5), 715–728.

Hyysalo, S (2003). Some problems in the traditional approaches of predicting the use of a technology-driven invention. *Innovation*, 16(2), 118–137.

Hyysalo, S (2004). *Uses of Innovation. Wristcare in the Practices of Engineers and Elderly*. Helsinki: Helsinki University Press.

Hyysalo, S (2006). The role of learning-by-using in the design of healthcare technologies: a case study. *The Information Society*, 22(2), 89–100.

Hyysalo, S and J Lehenkari (2003). An activity-theoretical method for studying user-participation in is design. *Methods of Information in Medicine*, 42(4), 398–405.

Hyysalo, S, J Lehenkari and R Miettinen (2003). Informaatiokumous, tuottaja-käyttäjäsuhteet ja sosiaaliset innovaatiot (Information revolution, user-producer relations and social innovations). In *Innovaatiopolitiikka*, T Lemola and P Honkanen (eds.), pp. 215–226. Helsinki: Gaudeamus.

Jaeger, B, R Slack and R Williams (2000). Europe experiments with multimedia: an overview of social experiments and trials. *The Information Society*, 16(4), 277–302.

Kalhama, M (2003). *Suomalaisen Peliteollisuuden Kartoitustutkimuksen Loppuraportti. (Charting Finnish Gaming Industry, Final Report)*. Helsinki: University of Art and Design Helsinki.

Kline, SJ and N Rosenberg (1986). An overview of innovation. In *The Positive Sum Strategy: Harnessing Technology for Economic Growth*, R Landau and N Rosenberg (eds.), pp. 275–305. Washington, DC: National Academy Press.

Laegran, AS and J Stewart (2003). Nerdy, trendy or healthy? Configuring the internet café. *New Media and Society*, 5, 357–377.

Latour, B (1987). *Science in Action: How to Follow Scientists and Engineers Through Society*. Cambridge, MA: Harvard University Press.

Latour, B (2005). *Reassembling the Social*. Oxford, UK: Oxford University Press.

Leadbeater, C (2006). *The User Innovation Revolution*. London, UK: Report by the National Consumer Council.

Lie, M and KH Sørensen (1997). Making technology our own?: domesticating technology into everyday life. In *Making Technology Our Own?: Domesticating Technology into Everyday Life*, M Lie and KH Sorensen (eds.), pp. 1–31. Oslo: Scandinavian University Press.

Lie, M and K Sørensen (eds.) (1996). *Making Technology Our Own? Domesticating Technology into Everyday Life*. Oslo: Scandinavian University Press.

Lundvall, B-Å (1988). Innovation as an interactive process: from user-producer interaction to the national system of innovation. In *Technical Change and Economic Theory*, G Dosi, C Freeman, RR Nelson, G Silverberg and L Soete (eds.), pp. 349–369. London: Pinter Publishers Ltd.

Lundvall, B-Å and B Johanson (1994). The learning economy. *Journal of Industry Studies*, 1(2), 23–42.

MacKenzie, D and J Wajcman (eds.) (1999). *The Social Shaping of Technology*, 2nd Ed. Buckingham: Open University Press.

MacKenzie, D (2006). Is economics performative? Option theory and the construction of derivatives markets. *Journal of the History of Economic Thought*, 28, 29–55.

McEvily, B and A Zaheer (1999). Bridging ties: a source of firm heterogeneity in conpetitive capabilities. *Strategic Management Journal*, 20, 1133–1156.

Miettinen, R (2002). *National Innovation System: Scientific Concept or Political Rhetoric*. Helsinki: Edita.

Miettinen, R, S Freeman, J Lehenkari, J Leminen, J Siltala, K Toikka and J Tuunainen (2008). Informaatiotekninen kumous, innovaatioverkostot ja luottamus (Information Revolution, Innovation Networks and Trust). Helsinki: Tekes.

Molina, A (1995). Sociotechnical constituencies as processes of alignment: the rise of a large-scale European information technology initiative. *Technology and Society*, 17(4).

Nelson, R and S Winter (1977). In search of useful theory of innovation. *Research Policy*, 6, 36–76.

Nicoll, DW (2000). Users as currency: technology and marketing trials as naturalistic environments. *The Information Society*, 16(4), 303–310.

Norman, D and S Draper (1986). *User Centered System Design: New Perspectives on Human-Computer Interaction*. Hillsdale, NJ: Lawrence Earlbaum.

Okamura, K, M Fujimoto, W Orlikowski and J Yates (1994). Helping CSCW applications succeed: the role of mediators in the context of use. In *Proc. of Computer Supported Collaborative Work Conference*. New York: ACM press, NC.

Osterloh, M and S Rota (2007). Open source software development — just another case of collective invention? *Research Policy*, 36(2), 157–171.

Pavitt, K (1984). Sectoral patterns of technical change: towards a taxonomy and a theory. *Research Policy*, 13, 343–373.

Pollock, N (2004). Universtiy or universality — on the establishment of the "organizationally generic". In *Proc. of the Paper Presented at "Understanding Sociotechnical Action" — Conference*. Edinburgh: Napier University (UK 3-4.7.2004).

Pollock, N and J Cornford (2004). ERP systems and the university as a "unique" organisation. *Information Technology and People*, 17, 31–52.

Pollock, N and R Williams (2008). *Software and Organizations: The Biography of the Packaged Enterprise System*. London: Routledge.

Pollock, N, R Williams and L D'Adderio (2007). Global software and its provenance: generification work in the production of organisational software packages. *Social Studies of Science*, 37(2), 254–280.

Preece, J, Y Rogers and H Sharp (2002). *Interaction Design — Beyond Human Computer Interaction*. Hoboken, NJ: John Wiley & Sons.

Procter, R and R Williams (1996). Beyond design: social learning and computer-supported cooperative work: some lessons from innovation studies. In *The Design of Computer-Supported Cooperative Work and Groupware Systems*, D Shapiro, M Tauber and R Traunmueller (eds.), Chap. 26, pp. 445–464 (1996). Amsterdam, The Netherlands: North Holland.

Rip, A, TJ Misa and J Schot (eds.) (1995). *Managing Technology in Society: The Approach of Constructive Technology Assessment*. London, NY: Pinter.

Rip, A and R Kemp (1998). Technological change. In *Human Choice and Climate Change*. Rayner, S and E Malone (eds.), pp. 327–399. Washington, D.C.: Battelle Press.

Rogers, EM (2003). *Diffusion of Innovations*. New York: The Free Press.

Rosenberg, N (1982). *Inside the Black Box: Technology and Economics*. Cambridge: Cambridge University Press.

Russell, S and R Williams (2002). Concepts, spaces and tools for action? Exploring the policy potential of the social shaping perspective. In *Shaping Technology, Guiding Policy: Concepts, Spaces and Tools*, K Sørensen and R Williams (eds.), pp. 133–154. Cheltenham, UK: Edward Elgar.

Silverstone, R and E Hirsch (1992). *Consuming Technologies: Media and Information in Domestic Spaces*. London: Routledge.

Sørensen, KH (1996). Learning technology, constructing culture. Socio-technical change as social learning. *STS Working Paper* no. 18/96, University of Trondheim: Centre for Technology and Society.

Stankiewicz, R (1995). The role of the science and technology infrastructure in the development and diffusion of industrial automation in Sweden. In *Technological Systems and Economic Performance: The Case of Factory Automation*, B Carlsson (ed.), pp. 165–210. Dordrecht: Kluwer.

Stewart, J (2004). Boys and girls stay in to play: creating computer entertainment for children, In *Private Sector Efforts to Include Women in ICTs*, C MacKeogh, P Preson (eds.), pp. 259–282. Trondheim: NTNU Working Paper Series.

Stewart, J (2007). Local experts in the domestication of ICTs. *Information Communication and Society*, 10(4), 547–569.

Stewart, J and R Williams (2005). The wrong trousers? Beyond the design fallacy: social learning and the user. In *User Involvement in Innovation Processes. Strategies and Limitations from a Socio-Technical Perspective*, H Rohracher (ed.), pp. 39–71. Munich: Profil-Verlag.

Stewart, J (2000). Cafematics: the cybercafe and the community. In *Community Informatics*. M Gurstein (ed.), pp. 320–338. Toronto: Idea Group.

Trigg, R and S Bodger (1994). From implementation to design: tailoring and the emergence of systematization in cscw. In *Proc. of Computer Supported Collaborative Work Conference*, pp. 45–54. New York: ACM press.

Van de Ven, AH, DE Polley, R Garud and S Venkataraman (1999). *The Innovation Journey*. Oxford: Oxford University Press.

von Hippel, E (1988). *The Sources of Innovation*. New York: Oxford University Press.

von Hippel, E (2005). *Democratizing Innovation*. Cambridge, MA: MIT Press.

von Hippel, E and M Tyre (1995). How learning by doing is done: problem identification in novel process equipment. *Research Policy*, 24(1), 1–12.

van Lieshout, M, T Egyedi and WE Bijker (eds.) (2001). *Social Learning Technologies: The Introduction of Multimedia in Education*. Aldershot: Ashgate.

Van der Meulen, B and A Arie Rip (1998). Mediation in the Dutch science system. *Research Policy*, 27, 757–769.
Williams, R and D Edge (1996). The social shaping of technology. *Research Policy*, 25, 856–899.
Williams, R and R Procter (1998). Trading places: a case study of the formation and deployment of computing expertise. In *Exploring Expertise*, R Williams *et al.* (eds.), pp. 197–222, Basingstoke: Macmillan.
Williams, R, R Slack and J Stewart (2005). *Social Learning in Technological Innovation — Experimenting with Information and Communication Technologies*. Cheltenham: Edgar Elgar Publishing.
Wolfe and Gertler (eds.) (2002). *Innovation and Social Learning: Institutional Adaptation in an Era of Technological Change* (pp. 260). Basingstoke, UK: Macmillan/Palgrave.

Part II

Drawing Users into the Innovation Process

Part I.

Drawing Lines into the Infinite and Beyond

USER-CENTRIC INNOVATIONS IN NEW PRODUCT DEVELOPMENT — SYSTEMATIC IDENTIFICATION OF LEAD USERS HARNESSING INTERACTIVE AND COLLABORATIVE ONLINE-TOOLS

VOLKER BILGRAM, ALEXANDER BREM
and KAI-INGO VOIGT

Friedrich-Alexander University of Erlangen-Nuremberg, Germany

Introduction

Innovations have long been considered to have a profound effect on the prosperity of businesses (Albach, 1989; Wheelwright and Clark, 1992; Cooper, 2002). However, their potential of growing into a competitive advantage coincides with an enormous failure-rate at the market especially in the field of breakthrough innovations (Crawford, 1997; Lüthje, 2007). Therefore, companies are trying to alleviate the lack of user-acceptance through opening their innovation processes to external actors, particularly customers (Brem, 2008). Such customer-centric innovations not only harness the voice-of-the-customer but also take a further step beyond the traditional market research by integrating users as problem solvers in various phases of the individual innovation process. In this context, the lead-user method is applied to capitalise on users with certain attributes, i.e., leading-edge customers, who are to benefit tremendously from innovative solutions. Hence, efficient processes and methods for a sustainable identification and integration of customers into the corporate innovation process are crucial to the success of new product development (NPD) (Herstatt, 1991; Olson and Bakke, 2001; Prügl, 2006; Brem and Voigt, 2007).

Another mega-trend of our time, trying to make use of the democratic powers of the Internet users, is epitomised by the buzzword web 2.0. After the "new economy" crash, the prevalent static Internet appearances of many dotcom businesses were gradually re-vitalised incorporating mechanisms that make use of the *wisdom of crowds* (O'Reilly, 2005). These web 2.0 applications such as online communities and weblogs are constantly getting more and more integrated in people's everyday lives — and meanwhile in companies' daily business as well.

The Internet and search engines in particular serve as a panecea for all kinds of search requests today. Yet, there is only a very limited body of research addressing the opportunities these new and highly personalised tools like communities and weblogs bear with respect to the efficient identification of lead users.

The primary aim of this paper is hence to elaborate the potential web 2.0 applications hold to support the systematic identification of lead users. This appears to be necessary in the light of the huge deficiencies the lead-user method that shows in the pivotal phase of the lead-user identification. Our approach was stimulated by a striking extensive networking of users and their willingness to reveal the personal as well as innovation-related information in online applications.

Literature Review

The way towards user-centric innovation was paved by the shift from the manufacturer-active to customer-active paradigm in the late 1970s (von Hippel, 1978a,b; Foxall and Tierney 1984). Since then, the development of user-centric innovation has constantly gained momentum and experienced a tremendous boost, in interest, in the wake of the widespread use of the Internet. The term, "lead user", was coined and conceptually developed by von Hippel (1986, 1988) more than 20 years ago. According to this concept, lead users are originally characterised by two fundamental criteria: first, they experience certain needs significantly earlier than the bulk of the market and thus serve as a "need-forecasting laboratory". Second, they are positioned to benefit notably from innovative solutions. For the purpose of practical applications of the method, a process consisting of four phases was devised (von Hippel, 1986; Lüthje and Herstatt, 2004): an initial preparatory phase and a phase of trend identification, i.e., trends the lead users are to be ahead of, are followed by the lead-user identification *per se*. In the frame of a workshop (phase 4), the identified lead users participate in the generation of new ideas and product concepts.

Since then, projects implementing the lead-user method were carried out in a variety of industries. This method proved to be a systematic approach for generating breakthrough innovations and was able to outperform comparable innovative approaches (Urban and von Hippel, 1988; Herstatt, 1991; von Hippel *et al.*, 1999; Lilien *et al.*, 2002; Morrison *et al.*, 2004). Based on the original lead-user concept, additional criteria and antecedents depicting lead users were explored, particularly facilitating the search for lead users in consumer goods markets with a substantially greater number of anonymous users as opposed to industrial markets (Lüthje, 2000; Lettl, 2004; Franke *et al.*, 2006).

Research also centred around search methods that are best suited to ensure an efficient identification of users characterised by the relevant search criteria (Urban and von Hippel, 1988; Prügl, 2006). Three distinct search methods have been conceived and applied: the *screening* method tests any person within a sample of all users for the presence of the criteria found relevant for the specific search purpose. The *pyramiding* or *networking* procedure takes a different approach starting out from a small number of persons and iteratively working its way up in the pyramid of expertise via recommendations. The *pyramiding* search distinguishes itself by increasing the chances of identifying leading-edge users in analogous markets following references of users in the target market. The *broadcasting* method has recently been applied in lead-user projects. Therein, the formulation of a problem is broadcasted to a group of potential problem solvers outside the company (Lakhani, 2006; Hienerth et al., 2007). However, research in this area indicates that companies are still facing considerable problems in efficiently identifying suitable users (Olson and Bakke, 2001; Lilien et al., 2002; Prügl, 2006; Brem and Voigt, 2007).

Surprisingly, the role of the Internet for the identification of lead users has scarcely been examined yet (Herstatt, 2003; Henkel and Sander, 2007). Different approaches were taken to integrate users in the various stages of the value chain, for instance, *mass customisation* or *toolkits for user innovation* (von Hippel, 2001; von Hippel and Katz, 2002; Thomke and von Hippel, 2002; Franke and von Hippel, 2003; Jeppesen, 2002; Prügl and Schreier, 2006), *community-based innovation* (Füller et al., 2006) or *netnography* (Kozinets, 1998, 2002; Füller et al., 2007). Whereas the core of *netnography* is to observe users' computer-mediated interaction, for instance, in communities, most other user-oriented innovation approaches aim at directly integrating users in the stage of idea generation and conceptualisation by shifting the individual trial-and-error process into an online realm by means of a user interface.

The principles of an *open innovation* (Chesbrough, 2003) and the co-operative innovative activities of users in communities have gained widespread notice alongside the evolvement of the Internet towards a platform connecting people and allowing for participation known under the notion Web 2.0 (O'Reilly, 2005). An extensive research on open source software (Raymond, 2001; von Hippel, 2001; Lerner and Tirole, 2002; Lakhani and von Hippel, 2003; von Hippel and von Krogh, 2003) and on innovative communities predominantly in the domain of extreme sports has been carried out. The probability of innovations to be generated, users' motives and willingness to share information are the aspects that have been examined as well as the characteristics of innovating users, their collaborative behaviour and the transformation of the character of the community (Shah, 2000; Franke and Shah, 2003; Lüthje, 2004; Hemetsberger and Reinhardt, 2004; Lüthje et al., 2005; Hienerth, 2006; Jeppesen and Frederiksen, 2006; Füller et al., 2006; Prügl and Schreier, 2006; Füller et al., 2007).

The following considerations will shed light on the lead-user method trying to conceptually link the recruitment of lead users to the online environment.

Conceptual Framework

The focus of this paper is the lead-user method, which can be delineated in the field of user-centred innovation concepts along multiple dimensions. In terms of a processual classification, the lead-user method is situated in the *fuzzy front end* (Khurana and Rosenthal, 1997; Kim and Wilemon, 2002) as a tool for systematic idea-generation and conceptualisation in the early phases of an innovation project. In contrast to the customer-specific configuration in later phases of the NPD, i.e., *mass customisation* using toolkits, the lead-user method does not limit the solution space within which users can generate ideas and is designed to integrate users in a face-to-face workshop rather than in an online setting. The non-representative nature is characteristic of the lead-user method that explicitly tries to explore the leading-edge customers' solutions to problems. Whereas traditional customer-oriented approaches concentrate on eliciting customers' representative needs in order to tailor their products to them, the lead-user method aims at users with exceptional qualities (Fig. 1).

In order to identify innovative user ideas with an outstanding commercial potential, the first criterion of a lead-user is that at the leading edge of significant trends in the market. Perceiving certain demands earlier than the bulk of the market — virtually living in the future as to a certain trend — enables lead users to gain

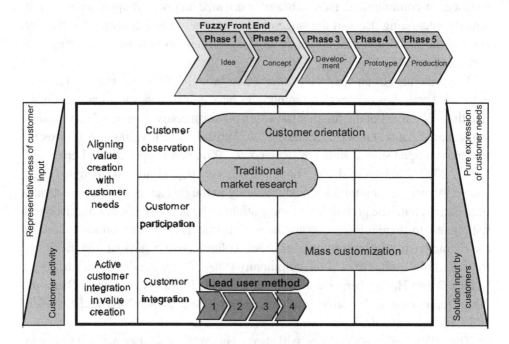

Fig. 1. The lead user method in the field of user-centred innovation.
Source: Wecht (2005); Verworn and Herstatt (2007).

"real-world experience" that is still further forward in time for the average consumer. Consequently, lead users are capable of overcoming an effect called *functional fixedness*, which explains the phenomenon that users' innovative potential is bound to the previous related experiences (Duncker, 1945; Birch and Rabinowitz, 1951; Adamson and Taylor, 1954; von Hippel, 1986). Although lead users are subject to the same cognitive restrictions, they are well set-up to create breakthrough innovations by virtue of their "living in the future". However, leading-edge needs do not necessarily entail the customer's motivation to innovate thus requiring a second component: users were found to be more inclined to innovate when they expect a high benefit from a solution to their needs (Mansfield, 1968; Urban and von Hippel, 1988; Franke et al., 2006).

Within the lead-user method, the identification of lead users according to the aforementioned criteria is of supreme importance, however, still showing room for improvements (Olson and Bakke, 2001; Lilien et al., 2002; Prügl, 2006). The process of identification can be divided into two main parts: 1. gathering appropriate criteria for the purpose of the innovation project and 2. screening users and identifying lead users who meet the set of criteria. The emphasis of this paper is laid on the compilation of the criteria used in the "offline world", which are to be accommodated to the online environment. Nevertheless, the following approach does not intend to draw a clear dividing line between the two parts of the identification process, but to integrate the pertinent aspects of the subsequent screening process in an effort to extend the application of criteria to the "online world".

Before we will commence the extension of the criteria, recent developments of the Internet are to be briefly outlined. The buzzword web 2.0 has been the subject of a plethora of discussions on the Internet that predominantly revolve around O'Reilly's deliberations (O'Reilly, 2005). Many attempts of making the notion web 2.0 understandable have certain fundamental principles and axioms in common among which is the participation of users in networked structures (O'Reilly, 2005; Kolbitsch and Maurer, 2006; Bienert, 2007; Maaß and Pietsch, 2007). This high degree of user interaction can be observed in an abundance of popular web 2.0 applications, e.g., MySpace, Facebook, Twitter, Wikipedia, Last.fm, YouTube, Flickr, Del.icio.us, Digg, Ning etc. Due to the collaborative and interactive essence of the Web 2.0 applications, the demarcation line between the producer and consumer has been notably blurred (Prahalad and Ramaswamy, 2000; Krempl, 2006; Bunz, 2006). Most prominent among the web 2.0 applications are the online communities, also known as social networking sites (Rheingold, 1993; Hagel and Armstrong, 1997; Whittaker et al., 1997; Preece, 2000; Prügl and Schreier, 2006; Füller et al., 2007; Henkel and Sander, 2007) and weblogs (Blood, 2000, 2004; Schmidt, 2006; Wright, 2006).

For the purpose of this paper, communities are defined as "social aggregations that emerge from the Net when enough people carry on public discussions long enough, with sufficient human feelings, to form webs of personal relationships in cyberspace" (Rheingold, 1993, S.5). Weblogs have generally been accepted to be frequently updated websites consisting of dated entries arranged in reverse chronological order (Blood, 2004; Efimova *et al.*, 2005; Schmidt *et al.*, 2005; Wright, 2006). The total number of weblogs called *blogosphere* is, sometimes, also considered a community (Efimova and de Moor, 2005; Efimova *et al.*, 2005). Likewise, the majority of the social networking sites (e.g., MySpace, Facebook, etc.) offer integrated weblogs as a core communication feature. Consequently, strict differentiation between the two applications is not practical.

As prior research made evident, users prefer innovating in groups rather than as isolated individuals (Franke and Shah, 2003; Lettl, 2004; Lüthje *et al.*, 2005; Füller *et al.*, 2007). Both online communities and weblogs enjoy increasing popularity and amount to a significant share of total media consumption (van Eimeren and Frees, 2006; Madden and Fox, 2006; Madden, 2006). Hence, Web 2.0 applications appear to be a key supplement to social "offline life" and consumption behaviour covering a huge variety of fields in terms of the content such as interests, hobbies or brands (Hagel and Armstrong, 1997; Kozinets, 1999; McWilliam, 2000; McAlexander *et al.*, 2002; Kozinets, 2002). The recent development of the online communities into platforms conducive to co-operative innovation activities asks for a re-consideration of the lead-user identification process.

Choosing the identification criteria as a starting point seems to be sound for two reasons. First, the criteria developed to identify lead users or innovative users have been devised in front of an offline background, thus lacking a conceptualisation tailored to the online environment. Second, the democratic nature of the web 2.0 may have the potential to overcome the anonymity prevalent in consumer markets that impedes a thorough identification of qualified users (Meffert, 1993; Wikström, 1996).

In an attempt to harness the innovative potential of users in communities and weblogs, two radically different methods can be adopted. Creating and establishing an individual online application solely designed to identify lead users (Spann and Skiera, 2003; Ernst *et al.*, 2004) can be juxtaposed with the usage of applications that already exist (Kozinets, 2002; Füller *et al.*, 2006). Within the framework of this paper, existing outside applications are to be concentrated on by virtue of the dominance and the enormous popularity they have already gained in the market. Leading-edge users are likely to be already committed to communities as active members. Thus, it appears to be difficult to induce them to join another network due to lock-in effects that communities have, once they have reached the critical mass of members.

A comprehensive analysis of the literature on user-centric innovation was accomplished for this paper and numerous search criteria from offline settings were compiled (Table 1). In addition to the original lead-user characteristics (von Hippel, 1988), criteria positively correlated to the lead-user construct and those describing

Table 1. Compilation of search criteria.

Criteria/*Indicator*	Sources
Lead user criteria	
Being ahead of a market trend	• von Hippel (1986) • Urban and von Hippel (1988) • von Hippel *et al.* (1999) • Herstatt (1991) • Lüthje (2000, 2004) • Franke *et al.* (2006) • Franke and Shah (2003) • Morrison *et al.* (2000) • Gruner (1997)
High expected benefit	• von Hippel (1986) • Urban and von Hippel (1988) • Herstatt (1991) • Franke and Shah (2003) • Franke *et al.* (2006) • Lettl (2004) • Lüthje (2004) • von Hippel and Riggs (1996) • Gruner (1997)
User investment	• von Hippel (1986) • Urban and von Hippel (1988) • Herstatt (1991) • Olson and Bakke (2001) • Nortel Networks (2000) • Stockmeyer (2001)
User dissatisfaction	• Urban and von Hippel (1988) • Herstatt (1991) • Lüthje (2000, 2004) • Olson and Bakke (2001) • Franke and Shah (2003) • Stockmeyer (2001)
Speed of adoption	• Urban and von Hippel (1988) • Herstatt (1991) • Olson and Bakke (2001) • Lüthje (2000, 2004) • Franke and Shah (2003) • Schreier and Prügl (2008) • Schreier *et al.* (2006)

Table 1. (*Continued*)

Criteria/*Indicator*	Sources
User Expertise	
Use experience	• Lüthje (2000, 2004) • Schreier and Prügl (2008) • Lüthje *et al.* (2005) • Franke and Shah (2003) • Hienerth *et al.* (2007)
Frequency of use	• Lüthje (2000) • Lettl *et al.* (2004) • Franke and Shah (2003) • Shah (2000) • Jeppesen and Frederiksen (2006) • Schreier and Prügl (2008) • Lettl (2004) • Franke *et al.* (2006)
Total period of use	• Lüthje (2000, 2004) • Franke and Shah (2003) • Schreier and Prügl (2008) • Lüthje *et al.* (2005)
Number of different disciplines	• Lüthje (2000, 2004) • Lettl (2004) • Lettl *et al.* (2004) • Lüthje *et al.* (2005)
Product related knowledge	• Lüthje (2000, 2004) • Jeppesen and Molin (2003) • Füller *et al.* (2006) • Gruner and Homburg (1998) • Schreier and Prügl (2008) • Lettl (2004) • Lüthje *et al.* (2005) • Hienerth *et al.* (2007)
Frequency of use of information sources	• Lüthje (2000, 2004) • Franke and Shah (2003) • Schreier and Prügl (2008)
Professional background or hobby	• Lüthje (2000, 2004) • Lüthje *et al.* (2005) • Herstatt (2003) • Jeppesen and Frederiksen (2006) • Hienerth *et al.* (2007)

Table 1. (*Continued*)

Criteria/*Indicator*	Sources
Motivation	
Extrinsic motivation	• Lüthje (2000, 2004) • Franke *et al.* (2006) • Lettl (2004)
Intrinsic motivation	• Lettl (2004) • Jeppesen and Frederiksen (2006) • Franke and Shah (2003) • Jeppesen and Molin (2003)
Extreme needs and circumstances of product use	• von Hippel *et al.* (1999) • von Hippel and Sonnack (1999) • Lettl (2004) • Herstatt *et al.* (2007) • Olson and Bakke (2004) • Lilien *et al.* (2002) • Lüthje *et al.* (2005) • Schild *et al.* (2004) • Herstatt (2001, 2003) • Nortel Networks (2000)
Opinion leadership and word-of-mouth	• Franke and Shah (2003) • Schreier and Prügl (2008) • Urban and von Hippel (1988) • Morrison *et al.* (2002) • Gatignon and Roberts (1985) • Sawhney an Prandelli (2000) • Lang (2006) • Lettl (2004) • Schreier *et al.* (2006) • Herstatt (2004)

innovative users have been taken into consideration as well. The collection of search criteria is meant to furnish a pool of criteria from which search criteria can be selected for one specific innovation project in order to correspond to a certain innovation purpose in the best possible way (Lüthje and Herstatt, 2004; Füller *et al.*, 2006; Hienerth *et al.*, 2007; Lüthje, 2007). In an effort to reveal the full potential of the web 2.0 applications for lead-user search, the set of offline criteria is "extrapolated" to the online environment of communities and weblogs. To accomplish this, the paper refers to the concepts in the science of social psychology and inter-personal relationships in computer-mediated networked environments. Moreover, specific features and structural characteristics of communities and weblogs are taken into

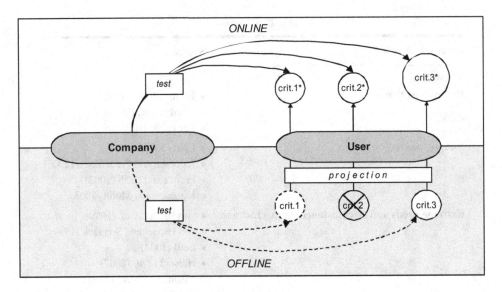

Fig. 2. Projection of user criteria.

consideration, as well as conspicuous new technologies in the field of web 2.0 applications. Three main paths of extrapolation are scrutinised, each revealing advantages in terms of an effective and efficient identification of lead users (Fig. 2): 1. certain criteria may be better tested in an online environment, i.e., usually rather anonymous consumers become more visible through an active participation in web 2.0 applications; 2. certain criteria might only be able to be assessed in the online setting, i.e., users feel inhibited to reveal certain characteristics in the real-world and 3. online users may show a higher degree of lead userness, i.e., users, who are active in web 2.0 applications, fulfill certain search criteria to a higher extent.

Conceptual Linkage–Lead-User Criteria and Web 2.0

In the following section, the potential of web 2.0 applications will be analysed with respect to each single criterion as structured in Table 1. For each criterion, we will provide theoretical background, elucidate its role in traditional lead-user identification and render arguments for an intensive integration of online applications into lead-user search.

Being ahead of a market trend

The first criterion to start our conceptual linkage is one of the two original lead-user criteria, *being ahead of a market trend*. Identifying users, who sense a certain need, before the majority of the market requires to investigate trends in an innovation field and identify the most critical ones. In a second step, users are screened for those leading these key trends (Herstatt and von Hippel, 1992; von Hippel et al., 1999). When projecting this criterion onto the online setting, we will concentrate on weblogs in particular, be they stand-alone applications or integrated features of communities.

Weblogs can be considered as the main communication tool in the web 2.0 context. For the identification of critical trends, a firm may benefit from technologies and structures of weblogs, which are conducive to trend identification in multiple ways: the strongly inter-linked nature of the *blogosphere* initially helps to single out weblogs with relevant content in a topical cluster by applying search engines that are specifically tailored to weblog search (Schmidt, 2006; Java, 2006; Mishne, 2006; Zerfaß and Bogosyan, 2007). Once a relevant set of weblogs is identified, the RSS technology allows for an easy and efficient surveillance by enabling users to subscribe to weblogs or sections of them and receive any updates immediately without having to navigate to a certain website (Hippner and Wilde, 2005; Bienert, 2007). Links from the observed set of weblogs to peripheral websites can serve as a relevance filter for a wide range of websites. This is facilitated by special features of the so-called permalinks, which unlike conventional hyperlinks, are bidirectional and offer more specified referral. Thereby, the reader can follow conversations that are spread out on several weblogs, as the blog entries refer to other entries (parts of a website), instead of just linking two websites (Glance et al., 2004; Hippner and Wilde, 2005; Schmidt, 2006; see also the *conversation tracker* by blogpulse). Discussions in weblogs are also encouraged by a technology called trackback that indicates at the end of a blog entry whether another weblog has referred to it. Furthermore, the spatial and temporal asynchronous weblog communication through the inherent comment feature at the end of each entry or the aforementioned trackback technology enables the leading-edge users to have an ongoing dialogue on trends without having to arrange a meeting. All of these communication features weblogs provide can help to ensure that a sufficient number of experts from analogous search fields, or those closely related, are taken into consideration for a comprehensive trend analysis (Lüthje, 2000; Pötz et al., 2005; Herstatt et al., 2007; Bienert, 2007).

Besides these weblog functions that are favourable to trend detection, the characteristic structure of weblogs shows potential to actually identify the users at the head of a market trend. Due to the reverse chronological order of the weblog entries and their allocation to categories within one weblog, a detected trend can easily be traced manually to its initiator. First approaches to automate the identification of certain types of bloggers have also been taken (Nakajima et al., 2005; Java, 2006).

Being an active participant in weblogs or communities might, as well, indicate a higher lead userness. The open nature and an easy usability of these web 2.0 applications turn them into assembly places of dynamic interdisciplinary and topical exchange of information and trends (Glance et al., 2004; Java, 2006; Kozinets, 2006;

Maurice, 2007). Additionally, the structure of weblogs is search-engine friendly, which entails high ranks in keyword search results, thus providing a breeding ground for analogous approaches to a problem beyond national boundaries or limitations to expert opinions of a certain branch of industry (Herstatt, 2003; Schmidt et al., 2005; Java, 2006; Schmidt, 2006).

High expected benefits

The second component of the original lead-user theory is the high benefit users expect from finding innovative solutions to a problem (Urban and von Hippel, 1988; Herstatt, 1991; Lüthje, 2000; Franke and Shah, 2003; Franke et al., 2006). Three indicators have been developed to allow for a better assessment of this criterion. Along these, the following projection of the second lead-user characteristic onto the online setting is structured.

User investment

In the absence of an adequate solution provided by a manufacturer, lead users will try to devise their own products or modify the existing products in order to satisfy their leading edge needs. These investments in obtaining a solution were found to be positively correlated with the benefit the user expects (Schmookler, 1966; Mansfield, 1968; von Hippel, 1988). However, *user investments* used to be perceptible only to a certain community, the user innovator belonged to. Therefore, this indicator could hardly be detected without conducting a survey (Ernst et al., 2004; Lüthje and Herstatt, 2004; Hienerth, 2006). Online communities have developed into platforms enabling users to reveal their ideas or innovations, for instance, by uploading drawings, virtual prototypes, CAD files or simply in discussions among the members. As a result, the indicator *user investment* can be more accurately and easily evaluated in an online setting.

By assessing specific visualisations of the user concepts and innovations as opposed to mere questioning of users, firms might be able to even reproduce and test user innovations (in contrast to Urban and von Hippel, 1988; Shah, 2004; Piller, 2006; Füller et al., 2007). Apart from the information on the user innovation, i.e., the *user investment*, the firm trying to evaluate the lead userness may also consider the opinion of other community members on the innovation, as it may be uttered in their comments. Hence, the firm's assessment might already be supplemented by a market perspective at this early stage of the NPD.

Not only lead users might be identified online in a more reliable and less complicated way, they might as well outdo innovative users in terms of the degree of expected benefit, which directly corresponds to a higher lead userness. On account of the collective effort of innovating within an online community, users may find favourable conditions for their innovative activities and as a result, the quality of *user investments* may tend to be superior (Amabile, 1997; Shah and Tripsas, 2004; Tietz et al., 2005; Hienerth, 2006; see also the discussion in the section "Use experience").

Füller et al. (2007) showed that there is often only a small number of innovative members in a community who play a key role in the collective generation of innovations. At the same time, these users proved to be widely known in the community and thus could easily be identified (Franke and Shah, 2003). Towards an automated process of identifying leading edge users, distinct user types could be recognised with respect to length, frequency and the level of postings they contribute within a thread (Henkel and Sander, 2007).

User dissatisfaction

The investment in finding a solution is cognitively preceded by a state of dissatisfaction with the existing products on the part of the user, which is the result of a negative discrepancy between the expected and the perceived performance of a product (Bruhn, 1982; Urban and von Hippel, 1988; Lüthje, 2000; Franke et al., 2006).

With regard to the indicator *user dissatisfaction*, potential of the web 2.0 can be found in the presumption that users would preferably express their discontent to a person in their peer group, e.g., another member of an online community which they are part of. The information ensuing from monitoring that kind of peer communication is said to be more reliable, unadulterated and unfiltered (Assael, 1998; Kozinets, 2002; Pitta and Fowler, 2005a). Online communities may also help to filter out postings from dissatisfied users that are not caused by mere ignorance or by a mistake of application, but rather reflect unfulfilled needs of leading-edge users. Utilising the self-organising effect of the communities' bottom-up structure is important, in order to be able to efficiently cope with the sheer multitude of user contributions in web 2.0 applications at all. It is plausible that by definition, lead users should only express dissatisfaction that originates in needs experienced way ahead of the bulk of the market rather than in a lack of *user experience*. Again, an automation based on the length of threads and the number of users involved in a discussion could be promising (Henkel and Sander, 2007).

The indicator *user dissatisfaction* is also supposed to ensure that users' general needs are transformed into specific product specifications (Lüthje, 2000). In discussions revolving around problem-solving, the community can assist specifying requirements and sustain *user dissatisfaction*.

Speed of adoption

Finally, the third indicator for the lead-user characteristic of *high expected benefit* is the *speed of adoption* regarding the new products. Research showed the higher the rate of adoption, the higher the benefit users expect from a new solution (Robertson, 1971; Urban and von Hippel, 1988; Rogers, 1995; Lüthje, 2004).

Again, monitoring communities or weblogs could replace the collecting of information by means of surveys. However, supposedly, the manually conducted surveillance of the web 2.0 applications is time-consuming in comparison to standardised questionnaires despite the focus on few communities and weblogs relevant to the innovation project, i.e., the search field. Users' reports of experience with products that were recently introduced to a market may signify their *speed of adoption* in the same way as announcements of a future intention to purchase or verbalised impatient expectations of new products.

It appears plausible that innovative users participating in web 2.0 applications show a tendency towards adopting new products faster, i.e., they fulfill the criterion to a larger extent. A possible reason can be seen in an earlier impetus for the decision-making process leading up to the adoption, i.e., purchase of a new product (Assael, 1998; Blackwell *et al.*, 2006). This, for instance, could happen when users gain awareness of a latent need due to the global communication and networking in online applications, e.g., when a product has already been introduced to a foreign market (Hennig-Thurau *et al.*, 2004; Füller *et al.*, 2007). The fast diffusion of information through weblogs and community communication and their trend-leading grassroots nature may accelerate the adoption (Gruhl *et al.*, 2005; Java, 2006). Subsequent phases of decision-making, such as the acquisition of product information and the evaluation of alternatives (Assael, 1998; Blackwell, 2006) can also gain momentum as a result of permanent involvement in online discussions and up-to-date reports of users' experiences (Gatignon and Robertson, 1985).

User expertise

Strikingly, both original lead-user characteristics rather concentrate on motivational qualities. The actual product-related abilities and the knowledge of users are not explicitly included in the original lead-user criteria. For this reason, several approaches have been taken to make users' expertise an integral part of lead-user theory. In doing so, a distinction is made in most researches between *user experience* and *product-related knowledge*. Both criteria have been found positively correlated with the lead-user construct and users' innovative activity (Lüthje, 2004; Lettl, 2004; Lüthje *et al.*, 2005).

In order to provide a theoretical framework for the extension of the lead-user criterion *user expertise* in an online environment, the section is preceded by a brief survey of theories. We are going to utilise the core statements of the three theories in the field of social psychology, which substantiate the processes of perception and learning (Lettl, 2004): the notion of *bounded rationality*, the theory of *social perception* and the conception of *absorptive capacity* will be outlined in the following with a focus on the elements that have relevance for our line of reasoning.

According to the notion of *bounded rationality*, individuals are not able to cope with all the information in a complex environment as a consequence of the restricted capacity and resources of the human memory. In order to obtain and assimilate information from the environment, individuals will reduce the complexity of the environment by concentrating on selective domains (Simon, 1957; Lipman, 1995; Selten, 2001; Dequech, 2001; Gigerenzer, 2001).

The theory of *social perception* postulates that the perception is determined by a set of hypotheses, which an individual has developed through experience, i.e., former perceptions and cognitions. If the hypotheses are confirmed by *user experience*, users will corroborate the set of hyphotheses. Otherwise, the perceptual set might be modified, i.e., the robustness of the hypotheses might be diminished. The set of hypotheses that is constantly re-considered and adapted also dominates the part of the environment that is perceived by the individual (Bruner and Postman, 1948; 1949; Lilli and Frey, 1993; Lettl, 2004).

The conception of *absorptive capacity* claims that the level of prior related knowledge, which an individual has accumulated, largely influences his/her ability to absorb information, evaluate it and utilise the knowledge for new solutions to a problem. As memory development is suggested to be self-reinforcing, an expanded knowledge base and the breadth of inter-linked categories are conducive to the assimilation and use of related knowledge (Bower and Hilgard, 1981; Cohen and Levinthal, 1990).

Apart from the theories in social psychology, another notion is used to explain the potential of web 2.0 applications in terms of the identification of lead users. The conception of *weak and strong ties* explores the social ties among the individuals in a network, e.g., an innovative online community. This conception may bear significance as far as the transfer of *user experience* and *product-related knowledge* between individual users is concerned. It may also affect the communication of relevant information in the buying decision of an individual user. The strength of social ties between individuals is gauged on the basis of several measures: the time invested to maintain social relationships, the emotional intensity and the degree of intimacy in relationships and on the basis of the level of mutual services. The conception therefore differentiates between weak and strong ties, which both have an effect on the behaviour of the individuals. Ties of different strength have also been argued to assume contrasting tasks regarding the functioning and structure of a network. For this reason, weak and strong ties are considered to be channels for the transfer of resources (Granovetter, 1973; Wasserman and Faust, 1994; Wegner, 1995). Information or resources that have been transferred via strong ties are acquired more easily and tend to be more detailed and reliable. This is particularly favourable in the context of complex pieces of information (Granovetter, 1983, 1985; Uzzi, 1997; Hansen, 1999). Weak ties, on the other hand, give an access to information beyond an individual's social structure, e.g., the individual is not member of a community (Granovetter, 1973, 1983; Johnson Brown and Reingen, 1987; Dodds et al., 2003; Jack, 2005; Kavanaugh et al., 2005).

Use experience

Use experience, as one component of *user expertise*, is developed by way of experiences from the repeated use of a product and therefore is primarily a matter of a user's time resources. It enables users to translate their dissatisfaction with solutions currently available on the market into specified requirements for the NPD by analysing problems and trialling new solutions (Weisberg, 1986; Alba and Hutchinson, 1987; Lüthje, 2000; 2004). The criterion *use experience* proved to be able to distinguish between innovating and non-innovating users and was found to positively correspond with lead userness (Franke and Shah, 2003; Lüthje, 2004; 2005).

In view of the extension of the criteria in the context of web 2.0 applications, the collective *use experience* should play a pivotal role. The individual in a community or weblog contributes to this experience based on a certain extent and is granted access in return by means of the interaction and communication within the network (Sawhney and Prandelli, 2000; Nambisan, 2002; Hienerth, 2006; Kolbitsch and Maurer, 2006). In the following, we will alter the angle from an individual to community-based when scrutinising the potential of the web 2.0. Particular attention will be directed toward any intensifying effects on the extent to which users fulfill the criteria of expertise.

Individuals have been found to preferably avail themselves of personal contacts as a source of information (Katz and Lazarsfeld, 1955; Allen, 1977; Kozinets, 1999; Cross et al., 2001; Godes et al., 2005). The chance to acquire tacit knowledge resulting from other users' time-consuming accumulation of *use experience* via trial-and-error may account for this behaviour (Shah, 2000; Lüthje et al., 2005; Tietz et al., 2005). Research on offline user communities, for instance, revealed that in 68% of all innovation projects at least three more members were involved besides the innovator. In 21% of all the cases, at least six other members were part of the team (Shah, 2000; Franke and Shah, 2003; Kozinets, 2006; similar Hienerth, 2006). In order to cover their need for information in various domains, users may take advantage of the resources of the community. Hence, users are required to activate social ties and divisions of their network according to their information endowment (Wellman and Gulia, 1999; von Hippel, 2005).

Making use of the collective experiences of a community may provide the means to overcome the cognitive limits, a single individual is subject to (Sawhney and

Prandelli, 2000; Butler *et al.*, 2002). In online communities, individuals share essential resources, be they of cognitive, emotional or material nature (McAlexander *et al.*, 2002). This, for instance, occurs when questions and experiences are posted and discussed online, which may be regarded as a "meta" trial-and-error process on the level of a community. Members of a community may not only test and optimise products that they have created themselves, but also innovations generated and revealed by other members (Shah and Tripsas, 2004; Hienerth, 2006; Piller, 2006; Füller *et al.*, 2007). These considerations can be substantiated by the theories in social psychology described beforehand.

The significance that might be ascribed to the collective assimilation, evaluation and utilisation of *use experience* can be underpinned by the conception of *absorptive capacity*. Owing to a deeper and more widely diversified knowledge base, i.e., a base of experiences, communities' *absorptive capacity* may by far surpass that of an individual (Cohen and Levinthal, 1990; Wegner, 1995; Harhoff *et al.*, 2003; Lüthje *et al.*, 2005). A community's focal interest can be assumed to provide a common knowledge base its members have and thus make it easier for the members to comprehend and exploit new and more specific knowledge, which other users contribute (Cohen and Levinthal, 1990). Due to the more intense interaction among the members of communities, which is facilitated by the community features such as the member profiles, electronic communication and structured forums, a system develops in which individuals increasingly develop the so-called transactive memory. This is knowledge members gain of other members' fields of expertise. The more extensive users' transactive memory is, the better their access to the community's knowledge pool will be (Wegner, 1987; Wegner *et al.*, 1991; Thompson and Fine, 1999; Brandon and Hollingshead, 2004). The knowledge resulting from *use experience* may, for instance, be transferred via an open source software and CAD-files, as could be observed in a kitesurfing community. Based on the information and expertise accumulated and revealed as CAD-files by the community members, other users are able to resume and further the tests and developments (Piller, 2006). This collective innovating effort allows users to substantially capitalise on other users' expertise that would not have been available to them otherwise, i.e., in an offline context (Preece and Maloney-Krichmar, 2003; de Valck, 2005).

The theory of *social perception* gives rise to the assumption that in contrast to an isolated individual, a radical re-orientation is more likely to take place in a community as a result of its larger set of hypotheses. Even though a single member's set of hypotheses may be exceptionally robust and thus control the individual's cognitive processing, another member characterised by a less robust perceptual set might just reject the hypothesis, once the experience is shared with the community. Any favourable outcome ensuing from that change of perspective can be expected to be disseminated within the community on account of the interaction among the community members. Consequently, the entire community's set of hypotheses will be re-adjusted and the innovative activities re-aligned with the latest findings. This characteristic may significantly increase the likelihood of breakthrough innovations to prevail and thereby make communities the preferred environment for lead users all the more.

If users focus on specific domains in a field of application by reason of reducing complexity — as claimed in the notion of *bounded rationality* — they are able to selectively obtain and assimilate information. At the same time, their decisions are pre-disposed to be sub-optimal. For the processing of information inside communities, an idealised scenario could be delineated that may, to some extent, resolve the cognitive dilemma: users may supplement the innovative approaches of other members with complementary information from their respective domains in a processed and aggregated form and make it accessible to the community again (Franke and Shah, 2003; Lettl, 2004; Lüthje et al., 2005; Piller, 2006). The technological possibility to permanently link single contributions (cp. Being ahead of a market trend) and archive them as connected units of knowledge appears to be conducive to this scenario. Besides, the information could be easily retrieved harnessing community search engines screening the archives (Hagel and Armstrong, 1997; de Valck, 2005; Dellarocas and Narayan, 2005; Tietz et al., 2005). This scenario would entail a higher level of information and thus higher lead userness among the community members.

On the whole, the interaction of users inside communities seems to have considerable potential to overcome the cognitive shortcomings as depicted in the theories of social psychology. Powerful communities, as a whole, can draw on significantly larger human resources of their members, which is particularly valuable in the context of *use experience* resulting from of laborious trial-and-error (Franke and Shah, 2003; Lüthje et al., 2005; Hienerth, 2006; Baldwin et al., 2006; Füller et al., 2007). Hence, it is not surprising that community-based internal resources could be found to have a positive effect on the likelihood that users innovate as well as on the commercial attractiveness of user innovations (Franke et al., 2006).

Means of communication are required in order to access a community's knowledge base comprising each member's expertise. (Wegner, 1987; Wegner et al., 1991; Thompson and Fine, 1999). We will now take a closer look at the channels of communication and try to identify those users in a community, who profit most from collective *use experience*. The following considerations are based on the conception of *weak and strong ties*.

The number of strong ties plays a pivotal role as far as a member's access to the collective knowledge is concerned (Wegner, 1987; Franke and Shah, 2003; Cross and Sproull, 2004; Jack, 2005; Schulz, 2006). *Use experience* is characterised by its high complexity as a result of the tacit knowledge that constitutes it and the related product knowledge required to underpin these experiences (Polanyi, 1958;

von Hippel, 1994; Nightingale, 1998). This complexity demands a strong social relationship among the parties of knowledge transfer (Hansen, 1999). Strong ties have further relevance regarding them as links into more distant social spheres of a user's network, i.e., second-degree relationships ("friends-of-friends") (Jack, 2005). The more strong ties members can draw on, the more *use experience* they will be able to acquire, which positively correlates with the user's lead userness (Franke *et al.*, 2006). The inter-relation between the strength of social ties and users' innovativeness is also supported by research in offline communities that showed that innovating users spend 32% more time with other members than non-innovating users (Franke and Shah, 2003).

In many cases, social relationships in online communities may be categorised as weak ties owing to the virtual way the relationship is initiated and cultivated (Kraut *et al.*, 1998; Wellman and Gulia, 1999; Andersson *et al.*, 2007). However, one has to bear in mind that some of the virtual relationships are the online counterparts to solid real-life relationships and, thus, are to be considered strong ties. Weak ties may also serve as bridges into socially distant clusters. However, in contrast to the strong ties, they provide direct relationships instead of forming a connection through a strong tie. Thereby, weak ties are more likely to produce non-redundant information as opposed to rather identical information obtainable from a usually very homogeneous group of strong ties (Burt, 1992; Cross and Sproull, 2004). One way to assess a user's network in terms of weak and strong ties could be the list of contacts, which is featured by most online communities. Nonetheless, these lists usually do not differentiate between the strengths of ties, which remain to be evaluated based on the degree of interaction between two members.

Studies on users' expertise sometimes refer to the frequency of use, the total period of use and the number of different disciplines as measures indicating *use experience* (Lüthje, 2004; Tietz *et al.*, 2005; Jeppesen and Frederiksen, 2006). These indicators may offer further perspectives to elaborate the potential of online lead-user identification.

Product-related knowledge

The second criterion constituting a user's expertise is the *product-related knowledge* that comprises a product's way of functioning as well as knowledge of material, processes and technology. It empowers the user to convert product requirements into preliminary solutions (Lüthje, 2000). Between the *product-related knowledge* and users' innovative activities as well as lead userness, a positive relationship has been detected (Lüthje, 2004).

Instead of relying on users' self-appraisal (cp. User investment) or easily accessible information, e.g., the academic degree of a user in surveys (Herstatt and von Hippel, 1992; Lettl, 2004; Hienerth *et al.*, 2007), firms can derive information from users' online activities. With the projection of this criterion into the online setting, the assessment of users' *product-related knowledge* is supposed to be notably facilitated. Users rendering assistance to other members of a community can be logically assumed to have greater *product-related knowledge* than those members who enlist advice. This assumption has been validated by research showing that 50% of the innovating users compared to only 10% of non-innovating users offered support to other members, while 40% of the supporting users were considered to have an expert knowledge (Franke and Shah, 2003). Users issuing newsletters or moderating discussion forums may be another sign of extensive inherent *product-related knowledge* (Seufert et al., 2002).

The frequency of use of information sources and the professional background or hobby could be identified as indicators of the extent of *product-related knowledge*, which are to be briefly discussed in the following. As far as the first indicator is concerned, similar collective effects and consequences should be present for *product-related knowledge* as have been discussed for the criterion *use experience* (cp. Use experience). In NPD, *product-related knowledge* in several domains is often a prerequisite for the innovative combination of a product's various components (Tietz et al., 2005; Piller, 2006). However, having profound expertise in several domains is rather unusual due to the long durations of the academic programmes. This is why research has predominantly focused on a user's professional background or hobby (Lüthje et al., 2002; Lüthje, 2004; Lüthje et al., 2005; Jeppesen and Frederiksen, 2006). Knowledge that is acquired within the bounds of one's profession or hobby is not involved in this dilemma as the local information ensuing from the pursuance of a profession or hobby is available to them at hardly any cost, i.e., the information is not "sticky". The low-cost procurement of information is owed to the fact that it is a by-product of the necessary pursuance of a profession or the enjoyable practice of a hobby (von Hippel, 1994; Morrison et al., 2000; Lüthje et al., 2005; Füller et al., 2007). This phenomenon is also consistent with the theory of *bounded rationality* stating that users focus on one or very few domains, particularly on one they are already proficient in, as a means to reduce the complexity (Simon, 1957; Selten, 2001; Dequech, 2001).

Research revealed that more than 70% of the innovating users had a profession they could transfer knowledge from for their innovative activities as opposed to 34% of non-innovating users (Lüthje, 2004). The example of a scientist working in the field of ergonomics and biomechanics, who used his/her professional knowledge to design a mountainbike frame in his/her spare time perfectly illustrates the effect of background knowledge on innovative activities (Lüthje et al., 2005).

In online communities "sticky" information can be easily transferred, i.e., made available to other members by way of visualisations that are uploaded (Ogawa, 1998; von Hippel, 1998). For this reason, online communities may have an edge over their offline equivalents that might reveal "sticky" information via face-to-face communication, however, only to a very limited number of members. A firm's online search for lead users with a background conducive to the innovation project may also be eased as a result of the extensive availability of innovation related information, i.e., in user profiles or forums, as well as advanced search capabilities in online communities (O'Murchu et al., 2004; Kolbitsch and Maurer, 2006; Kho, 2007; Lampe et al., 2007). The enormous revelation of information in online communities may further allow firms to select lead users by cumulatively combining users' expected benefits and backgrounds according to the respective search field. Provided that the search field of a lead-user project is described as "safe mountainbikes", the firm might search for users, who pursue exceptionally dangerous mountainbiking disciplines and at the same time are medical doctors (Lüthje et al., 2002; 2005).

User motivation

User motivation can be generally divided into extrinsic and intrinsic motivation.

Extrinsic motivation

Extrinsic motivation arises from the consequences of a user's activity and its attendant circumstances, i.e., monetary incentives or the benefit of using an innovation. These work as an impetus for innovative activities from the outside (Amabile, 1997; Frey, 1997; Lettl, 2004; Reichwald et al., 2004). The motivating power of benefits resulting from the utilisation of innovative products is already taken into account in the form of the second lead-user criterion *high expected benefit* (cp. High expected benefit). The effect of monetary incentives on users' motivation could not be confirmed in several studies. A possible explanation for this finding might be the suppression of intrinsic by extrinsic motivation (Herstatt and von Hippel, 1992; Lüthje, 2000; Jeppesen and Frederiksen, 2006; Pötz et al., 2005). For this reason, this kind of extrinsic motivation will not be further analysed.

In the context of communities, users' commitment to innovative activities of other members is evident. This being the case, social motives should be taken into consideration as the causes of extrinsic motivation. Users' behaviour is assumed to be under the influence of other members, so that the community develops a dynamic that may give a boost to users' motivation to innovate. Reasons for this could be the expectation of recognition or altruistic motives (Lerner and Tirole, 2002; Lakhani and von Hippel, 2003; Reichwald et al., 2004; Jeppesen and Frederiksen, 2006).

As can be seen from the example of an open-source development, user interaction and participation may have a strong motivational effect on users (Lerner and Tirole, 2002). This, however, is a phenomenon of the community as a whole, which cannot be transformed into a criterion on the level of a single user. Consequently, a user's membership of and commitment to a community can only serve as indicators of higher motivation *per se*, instead of allowing for differentiation between users' degree of motivation. Relevant communities may be assessed as to their effect on users' extrinsic social motivation in a first step. Then, users of favourable communities may only be further considered. (cp. Online commitment and participation as a pre-requisite of lead userness).

Intrinsic motivation

Intrinsic motivation results from the activitiy itself conveying a feeling of enjoyment, exploration and creativity to the users and enabling them to make full use of their potential (Zimbardo, 1992; Ryan and Deci, 2000; Reichwald *et al.*, 2004). The assumption that high intrinsic motivation has a positive influence on a user's innovative activities could only be partially confirmed in studies (Lüthje, 2000; Lettl, 2004; Lakhani, 2006; Jeppesen and Frederiksen, 2006). In the light of the analytical approach of this study, we will still extend this criterion to the online context. In contrast to the extrinsic motivation, intrinsic motivation can be examined on the individual level of a single user. In the context of web 2.0 applications, users' commitment could be assessed according to their participation to the community, i.e., revealing information on innovation projects or assisting other members (Franke and Shah, 2003; Füller *et al.*, 2007). It may help to segment a community population based on their commitment to the community and involvement in the topic of the community (Kozinets, 1999; 2002; Hemetsberger, 2001).

Moreover, individuals who, on the whole, firmly believe that certain outcomes are for the most part a result of their own actions, i.e., individuals with a high internal locus of control, tend to show high intrinsic motivation and creativity (Rotter, 1966; Leone and Burns, 2000). The construct of the locus of control is assumed to experience further reinforcement in the context of web 2.0 applications. These often have bottom-up organisational structures instead of being organised by a higher entity. Thus, users may increasingly embrace the role of a "producer" determining their own activities, i.e., the frequency and type of contributions (Hemetsberger, 2001; Bunz, 2006; Krempl, 2006; Kolbitsch and Maurer, 2006). This experience of far-reaching control might positively influence both a users' motivation and lead userness (Pitta and Fowler, 2005b).

Extreme needs and circumstances of product use

Extreme needs and conditions, a certain type of user may be confronted with, are rather a recommendation for a promising field to search in than search criteria themselves (von Hippel *et al.*, 1999; Nortel Networks, 2000; Herstatt, 2003; Lüthje and Herstatt, 2004; Schild *et al.*, 2004; Lettl, 2004; Lüthje *et al.*, 2005). Nevertheless, this recommendation will be extended to the online environment and examined in this context.

Expecting users to display a more intense and strongly motivated search behaviour under extreme conditions seems to be cogent. A higher pressure to find a solution to a problem due to the possible fatal consequences of product use may support this assumption as well as the users' open-mindedness towards new approaches. The latter is attributed to users in extreme situations, as they typically have a more open set of hypotheses and are therefore less impeded to innovate (Lettl, 2004).

Extreme users' intense search behaviour is likely to increase the probability of them tapping the resources and information offered by web 2.0 applications and searching for the few like-minded users that are also confronted with similar extreme conditions (Kolbitsch and Maurer, 2006; Schmidt, 2006; cp. Online commitment and participation as a pre-requisite of lead userness). Since extreme users will try to make use of a community's resources as much as possible, a higher *actor degree centrality* (Wasserman and Faust, 1994) and stronger commitment of users to a community might be further effects of the onerous external conditions that the extreme users face. Hence, frequent and informed contributions as well as a high number of social ties in a community should be the characteristic of extreme users.

Users with extreme needs may also be more easily or even solely identifiable in web 2.0 applications, since the problem they face might be extremely sensitive or stigmatising. In this case, users will rather prefer not to disclose any information despite their desperate search for a solution. Examples can be found in the field of diseases, disabilities and other sensitive subjects that require to recruit lead users from stigmatised segments of the society (Prügl, 2006). Under these conditions, communities as well as the weblogs enable users to interact and reveal information and at the same time remain anonymous. Still, it is feasible for a firm to gain valuable insights or even contact to extreme users via their online alias. This potential of an online lead-user search is crucial, as invaluable information can be obtained that otherwise would remain inaccessible.

Opinion leadership and word-of-mouth

Word-of-mouth is the informal communication of ideas, comments, opinions and information between two people, none of whom is a firm marketing its products (Godes *et al.*, 2005; Blackwell *et al.*, 2006). In the context of web 2.0, information is transmitted electronically by making it accessible online (Hennig-Thurau *et al.*, 2004). Opinion leaders are senders of information in a word-of-mouth process and are positioned to influence other individuals (de Valck, 2005; Anderrson *et al.*, 2007).

According to the conceptual framework, we focus on the identification and integration of lead users for the purpose of ideation and conceptualisation in the *fuzzy front end* of the innovation process (cp. Conceptual Framework). Despite this focus, we will briefly examine lead-users' ability to support corporate marketing of new products in the diffusion phase.

There are a variety of reasons why we attach importance to lead-users' opinion leadership. The marketing of breakthrough innovations that tend to be initially perceived as complex by customers, needs to offer assistance and guidance for customers in order to break with old habits of product usage (Atuahene-Gima, 1995; Veryzer, 1998). By virtue of their excellent *product-related knowledge* and *use experience*, lead users seem to be pre-destined for the role of opinion leaders (Urban and von Hippel, 1988; Schreier et al., 2006). Furthermore, the lead-user characteristics *being ahead of a market trend* and *speed of adoption* constitute the essence of opinion leadership (Rogers and Shoemaker, 1971; de Valck, 2005). By the time the majority of consumers sense a certain need, lead users will already have gained extensive experience and may accelerate the diffusion of a new product (von Hippel, 1986; Herstatt, 2004).

It becomes obvious that opinion leadership is already contained in the original lead-user criteria. This is also confirmed by the studies showing that lead userness and users' innovative activities are positively correlated to a user's opinion leadership (Franke and Shah, 2003; Morrison et al., 2004; Schreier et al., 2006). As a result, opinion leadership can be regarded as a by-product of lead userness.

Yet, we will point out a few additional criteria that might help to identify lead users or relevant communities (cp. Online commitment and participation as a prerequisite of lead userness) in the light of web 2.0 applications. First and foremost, we will concentrate on those criteria that derive from a user's membership of an online network and are generalisable rather than those that are tailored to a specific innovation project.

A large number of social contacts appear to be indispensable and essential to opinion leadership (Katz and Lazarsfeld, 1955). Communities can be considered networks that allow the informal transmission of information and consist of multiple relationships between their members (de Valck, 2005). The size of a user's network is decisive with respect to both dimensions, the total number of members and a user's direct contacts. Determining the size of offline networks is susceptible to mistakes as it is usually not publicly manifest and based on surveys (Katz and Lazarsfeld, 1955; Milgram, 1967). The size can be easily and more reliably determined in online networks based on the available media data that also include further details, such as page views, unique visitors or growth rates. On the level of individual users, direct contacts are often visible in each user's profile.

Additionally, online users might tend to have greater opinion leadership on account of the the common ground community members have and the larger number of contacts that can be reached at low cost (Moon and Sproull, 2001; Kozinets, 2002; de Valck, 2005).

In terms of the strength of social relationships among the members of a network, both weak and strong ties have different positive effects on the diffusion of information in the word-of-mouth process (Milgram, 1967; Granovetter, 1983; Constant *et al.*, 1996; Bickart and Schindler, 2001; Jack, 2005). Consequently, differentiation according to tie strength generally may not help to identify lead users.

Online commitment and participation as a pre-requisite of lead userness

The projection of search criteria into the web 2.0 as discussed in the previous sections showed that the identification of lead users yields advantages when the criteria are tested in the environment of a community or in the *blogosphere*. Thus, it might even be reasonable to go so far as to obligate a user to be a member of a community or to participate in a weblog as a pre-requisite of true lead userness. A less drastic course of action would be to set the starting point of lead-user search in the conducive environment of a community or the *blogosphere*. After all, the key question has to be posed: can there be true (high) lead userness without a user's commitment to or participation in an online application or should it become a constituent precondition for lead userness?

A possible line of reasoning in favour of the suggested incorporation of users' online-presence as a pre-condition of lead userness could be as follows: the users' high motivation is an essential part of lead-user theory and found expression in the component *high expected benefit* (cp. High expected benefit). It is claimed that a lead user, who is substantially dissatisfied with available products in the market, will tremendously invest in obtaining solutions to his/her strong needs (Urban and von Hippel, 1988; Franke *et al.*, 2006). If a user limits these search efforts to the offline environment, not drawing on the enormous resources, i.e., the support or collective knowledge of a community available online, it may be questioned whether true lead userness can be ascribed to that user.

Even though critical voices may claim that this pre-requisite entails a pre-selection of users that is too rigorous and restrictive, this concern is more and more resolved in view of the great and constantly increasing popularity of the online communities and weblogs among the innovative users. Nevertheless, choosing the online context as a starting point for lead-user search considerably restricts the search area. This is particularly true for the *screening* search method that does not allow references beyond the sample of users inspected, as opposed to the *pyramiding* search process (Prügl, 2006).

Acting on the suggestion to make online activity an integral part of lead-user theory, a two-fold process is necessary in which relevant communities or weblogs are selected first in order to identify lead users from these applications in a second step (Kozinets, 2002; Franke and Shah, 2003; Füller et al., 2006). In recent projects applying the lead-user method, a similar approach was taken. A search procedure called *broadcasting* was employed, selecting communities, forums or threads to "broadcast", i.e., formulate and reveal a problem to its users (Hienerth and Pötz, 2006; Hienerth et al., 2007). Depending on what level of the website structure (i.e., how deeply rooted), the problem is "broadcasted" (e.g., homepage, sub-categories, single threads), the sample of users to start with the search is wider or narrower.

As regards to the identification of relevant communities and weblogs, firms may either investigate the online landscape (cp. Being ahead of a market trend) or enquire relevant applications of experts and already identified lead users, harnessing the same networking technique used for the lead-user search (Prügl, 2006).

Most important of all to the selection of relevant online applications seems to be the specific search field of an innovation project (von Hippel et al., 1999; Herstatt et al., 2007). The topic or main interest of a community or weblog (e.g., a certain product or hobby) could serve as an aid to orientation and be an indicator of users' expertise. However, this topical selection of web 2.0 applications can only be carried out for each individual innovation project. Hence, in the following section, we will only take a brief look at criteria that are generalisable and may be considered for any lead-user project.

Research in communities showed that those in a less competitive environment tend to consist of users, who are more likely to assist other members and freely reveal information (Franke and Shah, 2003; von Hippel, 2002). With mutual assistance and free revealing mainly accounting for the superiority of online communities in comparison to single users or offline communities, it appears to be promising to opt for communities characterised by the limited competition or merely friendly rivalry (Jeppesen and Molin, 2003; Schulz, 2006; Hienerth, 2006; Jeppesen and Frederiksen, 2006; Füller et al., 2007).

Generally, there is a positive relationship between the size of a community and the efficiency of information search within the network (Baldwin et al., 2006; Hienerth and Pötz, 2006). This correlation can be explained by the powerful collective effects and the higher probability of identifying users with an exceptionally high lead userness in large communities.

User postings to a bulleting board with a minimum level of quality as well as a certain degree of interaction among members also seem to be valid demands ensuring that a community is frequently visited and intensely used by its members (Füller et al., 2007).

In a second step, lead users are identified in the selected online applications via search methods (Prügl, 2006). *Pyramiding* search might, for instance, be modified as to deliberately choosing a community member as the person to commence the search with. With respect to the *screening* search approach, it may help to select a group of people inside a community or sub-community for the initial sample of users that is to be screened.

Conclusion and Future Research

This analysis aimed to conceptionally elucidate the potential of web 2.0 applications for the identification of lead users and demonstrated varied starting points towards a utilisation of this potential. As a result, certain cases and scenarios could be identified in which online-situated search efforts might be superior to the offline equivalent.

With regard to the first path of extrapolation (cp. Conceptual framework; Fig. 2), potential has been discovered allowing for the better testing of criteria. Firstly, scenarios became obvious in which criteria could be tested more comprehensively. For instance, it was demonstrated how the criterion *user investment* could be supplemented by a qualitative component. Secondly, the assessment of the extent to which a user would fulfill a criterion can be assumed to be more reliable when tested in online applications. The users' expressions of dissatisfaction with existing products, for instance, were found to be rather unfiltered and genuine in online communication with their peers (Kozinets, 2002). Besides, instead of relying on users' self-appraisals, their *product-related knowledge* and investments in obtaining solutions can be evaluated based on the extent to which they offer assistance to other members (Franke and Shah, 2003) or reveal visualizations and data online (Shah, 2004; Piller, 2006; Füller *et al.*, 2007). Thirdly, criteria may also be more patent when put to test. Extensive information on hobbies and professional background in users' profiles makes their *product-related knowledge* very obvious to a firm (Kolbitsch and Maurer, 2006; Kho, 2007; Lampe *et al.*, 2007). Users publishing online newsletters or moderating relevant discussion forums might also indicate the *product-related knowledge* that they have accumulated (Seufert *et al.*, 2002). It stands to reason that all these advantages would enable firms to profile lead users in more detail and thus choose lead users that are particularly suited to a specific innovation project (Lüthje *et al.*, 2002; 2005; Hienerth *et al.*, 2007). The analysis also offered explanations why the search for lead users in web 2.0 applications might be able to significantly reduce the anonymity common in consumer markets. However, the analysis also implied that the qualitative gains are often accompanied by additional expenditure.

The theoretical evidence that certain criteria may solely be ascertainable in web 2.0 applications turned out to be of secondary importance. This scenario is likely to be found only in certain industries that meet very sensitive and extreme consumer needs, for instance, in the health-care sector. Users often feel more comfortable revealing sensitive information as anonymous members of a community (Prügl, 2006).

Due to the massive resources and collective knowledge of communities, scenarios could be delineated in which online users have a higher lead userness than the individuals using a product solely offline. This scenario appeared to be the most striking when testing users' expertise, i.e., *use experience* and *product-related knowledge* (Franke and Shah, 2003; Shah and Tripsas, 2004; Hienerth, 2006; Piller, 2006; Füller *et al.*, 2007).

On account of the aforementioned advantages of lead-user search in an online environment, firms may be encouraged to further extend search locations to web 2.0 applications. This may help the lead-user method to become established as a standard tool for the NPD.

A crucial limitation of this analysis is its entirely theoretical approach. The study is based on a scrutiny of search criteria used in the literature on user-centric innovation in order to identify and integrate innovative users. The potential of lead-user identification in the context of web 2.0 applications is inferred from the existing criteria that are projected into the online environment referring to various concepts. Future research will have to validate the potential in lead-user projects and examine the efficiency of online lead-user search. This is especially necessary considering the way of proceeding, which mosaicked individualities that were observed in specific web 2.0 applications, but may not be generalisable. The potential might thus vary depending on the industry or product category in which a firm's innovation project is situated in as well as on the number of relevant online applications in the respective search field (Lüthje, 2004). Furthermore, the users' willingness to freely reveal information and assist one another may differ depending on the life cycle phase of the community at the time of observation (Shah, 2000; Franke and Shah, 2003; Hienerth, 2006).

Additionally, the analysis came to the conclusion that future lead-user projects will have to seriously consider whether it might be necessary to integrate a user's online commitment and participation into the lead-user construct by making it a pre-requisite for true lead userness. This step would give rise to further questions revolving around the identification of the relevant web 2.0 applications. Hence, future research might start with a systematic analysis of lead-user projects that have used broadcast search, since this method has basically established users' online commitment as a pre-condition of lead userness.

References

Adamson, RE and DW Taylor (1954). Functional fixedness as related to elapsed time and to set. *Journal of Experimental Psychology*, 47(2), 122–126.

Alba, JW and WJ Hutchinson (1987). Dimensions of consumer expertise. *Journal of Consumer Research*, 13(4), 411–454.

Albach, H (1989). Innovationsstrategien zur Verbesserung der Wettbewerbsfähigkeit. *Zeitschrift für Betriebswirtschaft*, 59(12), 1338–1352.

Allen, TJ (1977). *Managing the Flow of Technology: Technology Transfer and the Dissemination of Technological Information within the R&D Organization.* Cambridge: MIT Press.

Amabile, TM (1997). Motivating creativity in organizations: on doing what you love and loving what you do. *California Management Review*, 40(1), 39–58.

Andersson, J, M Blomkvist and M Holmberg (2007). Blog marketing: a consumer perspective. In http://www.diva-portal.org/diva/getDocument?urn_nbn_se_hj_diva-891-1__fulltext.pdf, [28 March 2008].

Assael, H (1998). *Consumer Behavior and Marketing Action.* Cincinnati: South Western College Publishing.

Atuahene-Gima, K (1995). An exploratory analysis of the impact of market orientation on new product performance: a contingency approach. *Journal of Product Innovation Management*, 12(4), 275–293.

Baldwin, C, C Hienerth and E von Hippel (2006). How user innovations become commercial products: a theoretical investigation and case study. *Research Policy*, 35(9), 1291–1313.

Bickart, B and RM Schindler (2001). Internet forums as influential sources of consumer information. *Journal of Interactive Marketing*, 15(3), 31–40.

Bienert, J (2007). Web 2.0: Die demokratisierung des Internet. *Information Management & Consulting*, 22(1), 6–14.

Birch, HG and HJ Rabinowitz (1951). The negative effect of previous experience on productive thinking. *Journal of Experimental Psychology*, 41(2), 121–125.

Blackwell, RD, P Miniard and J Engel (2006). *Consumer Behavior.* Mason: South Western.

Blood, R (2000). Weblogs: A history and perspective. In: http://www.rebeccablood.net/essays/weblog_history.html [28 March 2008].

Blood, R (2004). How blogging software reshapes the online community. *Communications of the ACM*, 47(12), 53–55.

Bower, GH and ER Hilgard (1981). *Theories of Learning.* Englewood Cliffs: Prentice-Hall.

Brandon, DP and AB Hollingshead (2004). Transactive memory systems in organizations: matching tasks, expertise, and people. *Organization Science*, 15(6), 633–644.

Brem, A (2008). *The Boundaries of Innovation and Entrepreneurship: Conceptual Background and Essays on Selected Theoretical and Empirical Aspects.* Wiesbaden: Gabler.

Brem, A and K-I Voigt (2007). Innovation management in emerging technology ventures: the concept of an integrated idea management. *International Journal of Technology,*

Policy and Management, Special Issue on Technology Based Entrepreneurship and the Management of Knowledge Bases, 7(3), 304–321.

Bruhn, M (1982). *Konsumentenzufriedenheit und Beschwerden: Erklärungsansätze und Ergebnisse einer Empirischen Untersuchung in Ausgewählten Konsumbereichen*. Frankfurt a.M: Lang.

Bruner, JS and L Postman (1948). Symbolic value as an organizing factor in perception. *The Journal of Social Psychology*, 27(2), 203–208.

Bruner, JS and L Postman (1949). Perception, cognition, and behavior. *Journal of Personality*, 18(1), 14–31.

Bunz, M (2006). Wenn der Kunde handelt. *Brand Eins*, 8(4), 96–102.

Burt, R (1992). *Structural Holes: The Social Structure of Competition*. Cambridge: Harvard University Press.

Butler, B, L Sproull, S Kiesler and R Kraut (2002). Community effort in online groups: who does the work and why. In *Leadership at a Distance: Research in Technologically Supported Work*, SP Weisband (ed.), pp. 171–194. Mahwah: Lawrence Erlbaum.

Chesbrough, H (2003). *Open Innovation: The New Imperative for Creating and Profiting from Technology*. Boston: Harvard Business School Press.

Cohen, WM and DA Levinthal (1990). Absorptive capacity: a new perspective on learning and innovation. *Administrative Science Quarterly*, 35(1), 128–152.

Constant, D, L Sproull and S Kiesler (1996). The kindness of strangers: the usefulness of electronic weak ties for technical advice. *Organization Science*, 7(2), 119–135.

Cooper, RG (2002). *Winning at New Products: accelerating the Process from Idea to Launch*. Cambridge: Perseus Books.

Crawford, CM (1997). *New Products Management*. Homewood: Irwin.

Cross, R and L Sproull (2004). More than an answer: information relationships for actionable knowledge. *Organization Science*, 15(4), 446–462.

Cross, R, A Parker, L Prusak and SP Borgatti (2001). Knowing what we know: supporting knowledge creation and sharing in social networks. *Organizational Dynamics*, 30(2), 100–120.

de Valck, K (2005). *Virtual Communities of Consumption: Networks of Consumer Knowledge and Companionship*. Rotterdam: ERIM.

Dellarocas, C and R Narayan (2005). What motivates people to review a product online?: a study of the product specific antecedents of online movie ratings. University of Maryland, Working Paper.

Dequech, D (2001). Bounded rationality, institutions and uncertainty. *Journal of Economic Issues*, 35(4), 911–929.

Dodds, PS, R Muhamad and DJ Watts (2003). An experimental study of search in global social networks. *Science*, 301(8), 827–829.

Duncker, K (1945). On problem solving. *Psychological Monographs*, 58(5), Whole No. 270.

Efimova, L and A de Moor (2005). Beyond personal webpublishing: an exploratory study of conversational blogging practices. In *Proc. of the 38th Annual Hawaii International Conference on System Sciences*. USA: Big Island.

Efimova, L, S Hendrick and A Anjewierden (2005). Finding "the life between buildings": an approach for defining a weblog community. In *Internet Research 6.0: Internet Generations Conference*. Chicago: USA.

Ernst, E, JH Soll and M Spann (2004). Möglichkeiten der lead-user-identifikation in online-medien. In *Produktentwicklung mit virtuellen Communities*, C Herstatt and JG Sander (eds.), pp. 121–140. Wiesbaden: Gabler.

Foxall, GR and J Tierney (1984). From CAP 1 to CAP 2: user-initiated innovation from the user's point of view. *Management Decision*, 22(5), 3–15.

Franke, N and E von Hippel (2003). Satisfying heterogeneous user needs via innovation toolkits: the case of Apache security software. *Research Policy*, 32(7), 1199–1215.

Franke, N and S Shah (2003). How communities support innovative activities: an exploration of assistance and sharing among end-users. *Research Policy*, 32(1), 157–178.

Franke, N, E von Hippel and M Schreier (2006). Finding commercially attractive user innovations: a test of lead user theory. *Journal of Product Innovation Management*, 23(4), 301–315.

Frey, BS (1997). *Not Just for the Money: An Economic Theory of Personal Motivation*. Cheltenham: Edward Elgar Publishing.

Füller, J, G Jawecki and H Mühlbacher (2007). Innovation creation by online basketball communities. *Journal of Business Research*, 60(1), 60–71.

Füller, J, M Bartl, H Ernst and H Mühlbacher (2006). Community based innovation: how to integrate members of virtual communities into new product development. *Electronic Commerce Research*, 6(2), 57–73.

Gatignon, H and TS Robertson (1985). A propositional inventory for new diffusion research. *Journal of Consumer Research*, 11(4), 849–867.

Gigerenzer, G (2001). The adaptive toolbox. In *Bounded Rationality: the Adaptive Toolbox*, G Gigerenzer and R Selten (eds.), pp. 37–50. Cambridge: MIT Press.

Glance, NS, M Hurst and T Tomokiyo (2004). Blogpulse: automated trend discovery for weblogs. In *Proc. of the WWW 2004 Annual Workshop on the Weblogging Ecosystem: Aggregation, Analysis and Dynamics*. New York: USA.

Godes, D *et al.* (2005). The firm's management of social interactions. *Marketing Letters*, 16(3/4), 415–428.

Granovetter, M (1973). The strength of weak ties. *American Journal of Sociology*, 78(6), 1360–1380.

Granovetter, M (1983). The strength of weak ties: a network theory revisited. *Sociology Theory*, 1(1), 201–233.

Granovetter, M (1985). Economic action and social structure: the problem of embeddedness. *American Journal of Sociology*, 91(3), 481–510.

Gruhl, D, R Guha, R Kumar, J Novak and A Tomkins (2005). The predictive power of online chatter. In *Proc. of the 11th ACM SIGKDD International Conference on Knowledge Discovery in Data Mining*, 78–87. Chicago: USA.

Hagel, J and AG Armstrong (1997). *Net Gain: expanding Markets through Virtual Communities*. Boston: Harvard Business School Press.

Hansen, MT (1999). The search-transfer problem: the role of weak ties in sharing knowledge across organization subunits. *Administrative Science Quarterly*, 44(1), 82–111.

Harhoff, D, J Henkel and E von Hippel (2003). Profiting from voluntary information spillovers: how users benefit by freely revealing their innovations. *Research Policy*, 32(10), 1753–1769.

Hemetsberger, A (2001). Fostering cooperation on the internet: social exchange processes in innovative virtual consumer communities. In *Proc. of the Annual ACR Conference*. Texas: USA.

Hemetsberger, A and C Reinhardt (2004). Sharing and creating knowledge in open-source communities: the case of KDE. In *Proc. of the 5th European Conference on Organizational Knowledge, Learning, and Capabilities*. Innsbruck: Austria.

Henkel, J and JG Sander (2007). Identifikation innovativer Nutzer in virtuellen communities. In *Management der frühen Innovationsphasen: Grundlagen — Methoden — Neue Ansätze*, C Herstatt and B Verworn (eds.), pp. 77–107. Wiesbaden: Gabler.

Hennig-Thurau, T, KP Gwinner, G Walsh and DD Gremler (2004). Electronic word-of-mouth via consumer-opinion platforms: what motivates consumers to articulate themselves on the internet. *Journal of Interactive Marketing*, 18(1), 38–52.

Herstatt, C (1991). *Anwender als Quellen für die Produktinnovation*. Zürich: ADAG Administration & Druck AG.

Herstatt, C (2003). Onlinegestützte Suche nach innovativen Anwendern in direkten und analogen Anwendermärkten. Technische Universität Hamburg-Haburg, Working Paper No. 21.

Herstatt, C (2004). Dialog mit Kunden und lead-user-management in der innovationspraxis. In *Das innovative Unternehmen: Produkte, Prozesse, Dienstleistungen*, H Barske, A Gerybadze, C Hünninghausen and T Sommerlatte (eds.), Wiesbaden: Symposion Publishing.

Herstatt, C and E von Hippel (1992). From experience: developing new product concepts via the lead user method: a case study in a "low tech" field. *Journal of Product Innovation Management*, 9(3), 213–221.

Herstatt, C, C Lüthje and C Lettl (2007). Fortschrittliche Kunden zu breakthrough-innovationen stimulieren. In *Management der frühen Innovationsphasen: Grundlagen — Methoden — Neue Ansätze*, C Herstatt and B Verworn (eds.), pp. 61–75. Wiesbaden: Gabler.

Hienerth, C (2006). The commercialization of user innovations: the development of the rodeo kayaking industry. *R&D Management*, 36(3), 273–294.

Hienerth, C and M Pötz (2006). Making the lead user idea-generation process a standard tool for new product development. In *Proc. of the 4th International Workshop on User Innovation*. Munich: Germany.

Hienerth, C, M Pötz and E von Hippel (2007). Exploring key characteristics of lead user workshop participants: who contributes best to the generation of truly novel solutions? In *Proc. of the DRUID Summer Conference 2007 on Appropriability, Proximity, Routines and Innovation*. Copenhagen: Denmark.

Hippner, H and T Wilde (2005). Social software. *Wirtschaftsinformatik*, 47(6), 441–444.

Jack, SL (2005). The role, use and activation of strong and weak network ties: a qualitative analysis. *Journal of Management Studies*, 42(6), 1233–1259.

Java, A (2006). Tracking influence and opinions in social media. In http://ebiquity.umbc.edu/_file_directory_/resources/203.pdf, [28 March 2008].

Jeppesen, LB (2003). The implications of "user toolkits for innovation". Copenhagen Business School, Working Paper 2002-09.

Jeppesen, LB and L Frederiksen (2006). Why do users contribute to firm-hosted user communities? The case of computer-controlled music instruments. *Organization Science*, 17(1), 45–63.

Jeppesen, LB and MJ Molin (2003). Consumers as co-developers: learning and innovation outside the firm. *Technology Analysis & Strategic Management*, 15(3), 363–383.

Johnson Brown, J and PH Reingen (1987). Social ties and word-of-mouth referral behavior. *Journal of Consumer Research*, 14(3), 350–362.

Katz, E and PF Lazarsfeld (1955). *Personal Influence: The Part Played by People in the Flow of Mass Communication*. New York: Free Press.

Kavanaugh, A, JM Carroll, MB Rosson, TT Zin and DD Reese (2005). Community networks: where offline communities meet online. In http://jcmc.indiana.edu/vol10/issue4/kavanaugh.html [28 March 2008].

Kho, ND (2007). Networking opportunities: social networking for business. *EContent*, 30(4), 24–29.

Khurana, A and SR Rosenthal (1997). Integrating the fuzzy front end of new product development. *MIT Sloan Management Review*, 38(2), 103–120.

Kim, J and D Wilemon (2002). Focusing the fuzzy front-end in new product development. *R&D Management*, 32(4), 269–279.

Kolbitsch, J and H Maurer (2006). The transformation of the web: how emerging communities shape the information we consume. *Journal of Universal Computer Science*, 12(2), 187–213.

Kozinets, RV (1998). On netnography: initial reflections on consumer research investigations of cyberculture. *Advances in Consumer Research*, 25(1), 366–371.

Kozinets, RV (1999). E-tribalized marketing: the strategic implications of virtual communities of consumption. *European Management Journal*, 17(3), 252–264.

Kozinets, RV (2002). The field behind the screen: using netnography for marketing research in online communities. *Journal of Marketing Research*, 39(1), 61–72.

Kozinets, RV (2006). Click to connect: netnography and tribal advertising. *Journal of Advertising Research*, 46(3), 279–288.

Kraut, R, M Patterson, V Lundmark, S Kiesler, T Mukopadhyay and W Scherlis (1998). Internet paradox: a social technology that reduces social involvement and psychological well-being? *American Psychologist*, 53(9), 1017–1031.

Krempl, S (2006). Soziale software: innovative Bausteine für eine kritische Netzöffentlichkeit. In *Die wunderbare Wissensvermehrung: Wie Open Innovation unsere Welt revolutioniert*, O Drossou, S Krempl and A Poltermann (eds.), pp. 168–180. Hannover: Heise.

Lakhani, K (2006). Broadcast search in problem solving: attracting solutions from the periphery. MIT Sloan School of Management, Working Paper.

Lakhani, KR and E von Hippel (2003). How open source software works: "free" user-to-user assistance. *Research Policy*, 32(6), 923–943.

Lampe, C, N Ellison and C Steinfield (2007). A familiar face (book): Profile elements as signals in an online social network. In *Proc. of the CHI 2007 Online Representation of Self*, 12–14. San Jose: USA.

Leone, C and J Burns (2000). The measurement of locus of control: assessing more than meets the eyes? *Journal of Psychology*, 134(1), 63–76.

Lerner, J and J Tirole (2002). Some simple economics of open source. *Journal of Industrial Economics*, 50(2), 197–234.

Lettl, C (2004). *Die Rolle von Anwendern bei hochgradigen Innovationen: Eine explorative Fallstudienanalyse in der Medizintechnik*. Wiesbaden: Deutscher Universitäts-Verlag

Lilien, GL, PD Morrison, K Searls, M Sonnack and E von Hippel, E (2002). Performance assessment of the lead user idea-generation process for new product development. *Management Science*, 48(8), 1042–1059.

Lilli, W and D Frey (1993). Die Hypothesentheorie der sozialen Wahrnehmung. In *Theorien der Sozialpsychologie*, D Frey and M Irle (eds.), pp. 49–78. Bern: Verlag Hans Huber.

Lipman, BL (1995). Information processing and bounded rationality: a survey. *Canadian Journal of Economics*, 28(1), 42–67.

Lüthje, C (2000). *Kundenorientierung im Innovationsprozess: Eine Untersuchung der Kunden-Hersteller-Interaktion in Konsumgütermärkten*. Wiesbaden: Gabler-Verlag.

Lüthje, C (2004). Characteristics of innovating users in a consumer goods field: an empirical study of sport-related product consumers. *Technovation*, 24(9), 683–695.

Lüthje, C (2007). Methoden zur Sicherstellung von Kundenorientierung in den frühen Phasen des Innovationsprozesses. In *Management der frühen Innovationsphasen: Grundlagen — Methoden — Neue Ansätze*, C Herstatt and B Verworn (eds.), pp. 39–60. Wiesbaden: Gabler.

Lüthje, C and C Herstatt (2004). The lead user method: an outline of empirical findings and issues for future research. *R&D Management*, 34(5), 553–568.

Lüthje, C, C Herstatt and E von Hippel (2002). The dominant role of "local" information in user innovation: the case of mountain biking. MIT Sloan School of Management, Working Paper, 4377–02.

Lüthje, C, C Herstatt and E von Hippel (2005). User-innovators and "local" information: the case of mountain biking. *Research Policy*, 34(6), 951–965.

Maaß, C and G Pietsch (2007). Web 2.0 als Mythos, Symbol und Erwartung. In *Diskussionsbeitrag der Fakultät für Wirtschaftswissenschaft der FernUniversität in Hagen* 408.

Madden, M (2006). Internet penetration and impact. In http://www.pewinternet.org/pdfs/PIP_Internet_Impact.pdf, [28 March 2008].

Madden, M and S Fox (2006). Riding the waves of web 2.0. In http://www.pewinternet.org/pdfs/PIP_Web_2.0.pdf, [28 March 2008].

Mansfield, E (1968). *Industrial Research and Technological Innovation: An Econometric Analysis*. New York: W.W. Norton.

Maurice, F (2007). *Web 2.0 Praxis: AJAX, Newsfeeds, Blogs, Microformats*. München: Markt + Technik.

McAlexander, JH, JW Schouten and HF Koenig (2002). Building brand community. *Journal of Marketing*, 66(1), 38–54.

McWilliam, G (2000). Building stronger brands through online communities. *Sloan Management Review*, 41(13), 43–54.

Meffert, H (1993). Konsumgütermarketing. In *Handwörterbuch der Betriebswirtschaft: Teilband 2*, W Wittmann, W Kern, R Köhler, HU Küpper and K von Wysocki (eds.), col. 2241–2255. Stuttgart: Schäffer-Poeschel.

Milgram, S (1967). The small-world problem. *Psychology Today*, 1(1), 60–67.

Mishne, G (2006). Predicting movie sales from blogger sentiment. In *Proc. of the AAAI 2006 Spring Symposium on Computational Approaches to Analysing Weblogs*. Palo Alto: USA.

Moon, JY and L Sproull (2001). Turning love into money: how some firms may profit from voluntary electronic customer communities. In http://userinnovation.mit.edu/papers/Vol-Customers.pdf, [28 March 2008].

Morrison, P, J Roberts and D Midgley (2004). The nature of lead user and measurement of leading edge status. *Research Policy*, 33(2), 351–362.

Morrison, PD and JH Roberts (2000). Determinants of user innovation and innovation sharing in a local market. *Management Science*, 46(12), 1513–1527.

Nakajima, S, J Tatemura, Y Hino, Y Hara and K Tanaka (2005). Discovering important bloggers based on analyzing blog threads. In *Proc. of the WWW 2005 2nd Annual Workshop on the Weblogging Ecosystem: Aggregation, Analysis and Dynamics*. Chiba: Japan.

Nambisan, S (2002). Designing virtual customer environments for new product development: toward a theory. *Academy of Management Review*, 27(3), 392–413.

Nightingale, P (1998). A cognitive model of innovation. *Research Policy*, 27(7), 689–709.

Nortel Networks (2000). Lead users and dynamic information transfer. In http://www.opencaseware.org/documents/nortelleaduser.pdf, [28 March 2008].

O'Murchu, I, JG Breslin and S Decker (2004). Online social and business networking communities. In *Proc. of the 16th ECAI, Workshop on the Application of Semantic Web Technologies to Web Communities*. Valencia: Spain.

O'Reilly, T (2005). What is web 2.0: design patterns and business models for the next generation of software. In http://www.oreilly.com/pub/a/oreilly/tim/news/2005/09/30/what-is-web-20.html, [28 March 2008].

Ogawa, S (1998). Does sticky information affect the locus of innovation? Evidence from the Japanese convenience-store industry. *Research Policy*, 26(7/8), 777–790.

Olson, EL and G Bakke (2001). Implementing the lead user method in a high technology firm: a longitudinal study of intentions versus actions. *Journal of Product Innovation Management*, 18(6), 388–395.

Piller, FT (2006). User Innovation: Der Kunde kann's besser. In *Die wunderbare Wissensvermehrung: Wie Open Innovation unsere Welt revolutioniert*, O Drossou, S Krempl and A Poltermann (eds.), pp. 85–97. Hannover: Heise.

Pitta, DA and D Fowler (2005a). Internet community forums: an untapped resource for consumer marketers. *Journal of Consumer Marketing*, 22(5), 265–274.

Pitta, DA and D Fowler (2005b). Online consumer communities and their value to new product developers. *Journal of Product & Brand Management*, 14(5), 283–291.

Polanyi, M (1958). *Personal Knowledge: Towards a Post-Critical Philosophy*. Chicago: University of Chicago Press.

Pötz, M, C Steger, I Mayer and J Schrampf (2005). Evaluierung von Case Studies zur Lead User Methode. In http://www.iluma.at/pdf/Ergebnisse/WP340%20Evaluierung%20 Case%20Studies.PDF, [28 March 2008].

Prahalad, CK and V Ramaswamy (2000). Co-opting customer competence. *Harvard Business Review*, 78(1), 79–87.

Preece, J (2000). *Online Communities: Designing Usability, Supporting Sociability*. Chichester: Wiley.

Preece, J and D Maloney-Krichmar (2003). Online communities: focusing on sociability and usability. In *Handbook of Human-Computer Interaction*, J Jacko and A Sears (eds.), pp. 596–620. Mahwah: Lawrence Erlbaum Associates.

Prügl, R (2006). Die Identifikation von Personen mit besonderen Merkmalen: Eine empirische Analyse zur Effizienz der Suchmethode Pyramiding. Doctoral dissertation, Wirtschafsuniversität Wien: Department of Entrepreneurship and Innovation.

Prügl, R and M Schreier (2006). Learning from leading-edge customers at the sims: opening up the innovation process using toolkits. *R&D Management*, 36(3), 237–250.

Raymond, E (2001). *The Cathedral and the Bazaar: Musings on Linux and Open Source by an Accidental Revolutionary*. Sebastopol: O'Reilly.

Reichwald, R, C Ihl and S Seifert (2004). Innovation durch Kundenintegration. *Lehrstuhls für Allgemeine und Industrielle Betriebswirtschaftslehre an der Technischen Universität München, Arbeitsbericht* 40.

Rheingold, H (1993). *The Virtual Community: Homesteading on the Electronic Frontier*. Reading: Addison-Wesley.

Robertson, TS (1971). *Innovative Behavior and Communication*. New York: Holt Rinehart and Winston.

Rogers, EM (1995). *Diffusion of Innovations*. New York: Free Press.

Rogers, EM and FF Shoemaker (1971). *Communication of Innovations: A Cross-Cultural Approach*. New York: Free Press.

Rotter, JB (1966). Internal-external control and reinforcement. *Psychological Monographs*, 80(1), 609.

Ryan, RM and EL Deci (2000). Intrinsic and extrinsic motivations: classic definitions and new directions. *Contemporary Educational Psychology*, 25(1), 54–67.

Sawhney, M and E Prandelli (2000). Beyond customer knowledge management: customers as knowledge co-creators. In *Knowledge Management and Virtual Organizations*, Y Malhotra (ed.), pp. 258–281. Hershey: Idea Group Publishing.

Schild, K, C Herstatt and C Lüthje (2004). How to use analogies for breakthrough innovations. Technische Universität Hamburg-Haburg, Working Paper 24.

Schmidt, J (2006). *Weblogs: Eine kommunikationssoziologische Studie.* Konstanz: UVK Verlagsgesellschaft.

Schmidt, J, K Schönberger and C Stegbauer (2005). Erkundungen von Weblog-Nutzungen: Anmerkungen zum Stand der Forschung. In *Erkundungen des Bloggens: Sozialwissenschaftliche Ansätze und Perspektiven der Weblogforschung*, J Schmidt, K Schönberger and C Stegbauer (eds.), pp. 1–19, Special edition of kommunikation@gesellschaft, Vol. 6, Online publication.

Schmookler, J (1966). *Invention and Economic Growth.* Cambridge: Harvard University Press.

Schreier, M and R Prügl (2008). Extending lead user theory: antecedents and consequences of consumers' lead userness. *Journal of Product Innovation Management*, 25(4), 331–346.

Schreier, M, S Oberhauser and R Prügl (2006). Lead users and the adoption and diffusion of new products: insights from two extreme sports communities. *Marketing Letters*, 18(1/2), 15–30.

Schulz, C (2006). The secret to successful user communities: an analysis of computer associates' user groups. Munich School of Management, Working Paper 2006-13.

Selten, R (2001). What is bounded rationality? In *Bounded Rationality: The Adaptive Toolbox*, G Gigerenzer and R Selten (eds.), pp. 13–36. Cambridge: MIT Press.

Seufert, S, M Moisseeva and R Steinbeck (2002). Virtuelle Communities gestalten. In *Handbuch E-Learning*, A Hohenstein and K Wilbers (eds.), pp. 1–20. Köln: Fachverlag Deutscher Wirtschaftsdienst.

Shah, S (2000). Sources and patterns of innovation in a consumer products field: innovations in sporting equipment. MIT Sloan School of Management, Working Paper 4105.

Shah, SK (2004). From innovation to firm and industry formation in the windsurfing, skateboarding and snowboarding industries. University of Illinois, Working Paper 05-0107.

Shah, SK and M Tripsas (2004). When do user-innovators start firms?: towards a theory of user entrepreneurship. University of Illinois, Working Paper 04-0106.

Simon, HA (1957). *Administrative Behavior: A Study of Decision-Making Processes in Administrative Organization.* New York: Macmillan.

Spann, M and B Skiera (2003). Internet-based virtual stock markets for business forecasting. *Management Science*, 49(10), 1310–1326.

Thomke, S and E von Hippel (2002). Customers as innovators: a new way to create value. *Harvard Business Review*, 80(4), 74–81.

Thompson, L and GA Fine (1999). Socially shared cognition, affect, and behavior: a review and integration. *Personality and Social Psychology Review*, 3(4), 278–302.

Tietz, R, PD Morrison, C Lüthje and C Herstatt (2005). The process of user-innovation: a case study in a consumer goods setting. *International Journal of Product Development*, 2(4), 321–338.

Urban, GL and E von Hippel (1988). Lead user analysis for the development of new industrial products. *Management Science*, 34(5), 569–582.

Uzzi, B (1997). Social structure and competition in interfirm networks: the paradox of embeddedness. *Administrative Science Quarterly*, 42(1), 35–67.

van Eimeren, B and B Frees (2006). ARD/ZDF-Online-Studie 2006: Schnelle Zugänge, neue Anwendungen, neue Nutzer? *Media Perspektiven*, 8, 402–415.

Verworn, B and C Herstatt (2007). Bedeutung und Charakteristika der frühen Phasen des Innovationsprozesses. In *Management der frühen Innovationsphasen: Grundlagen — Methoden — Neue Ansätze*, C Herstatt and B Verworn (eds.), pp. 3–19. Wiesbaden: Gabler.

Veryzer, RW (1998). Key factors affecting customer evaluation of discontinuous new products. *Journal of Product Innovation Management*, 15(2), 136–150.

von Hippel, E (1978a). A customer-active paradigm for industrial product idea generation, *Research Policy*, 7(3), 240–266.

von Hippel, E (1978b). Successful industrial products from customer ideas. *Journal of Marketing*, 42(1), 39–49.

von Hippel, E (1986). Lead users: a source of novel product concepts. *Management Science*, 32(7), 791–805.

von Hippel, E (1988). *The Sources of Innovation*. New York: Oxford University Press.

von Hippel, E (1994). Sticky information and the locus of problem solving: implications for innovation. *Management Science*, 40(4), 429–440.

von Hippel, E (1998). Economics of product development by users: impact of sticky local information. *Management Science*, 44(5), 629–644.

von Hippel, E (2001). Innovation by user communities: learning from open-source software. *MIT Sloan Management Review*, 42(4), 82–86.

von Hippel, E (2002). Horizontal innovation networks: by and for users. MIT Sloan School of Management, Working Paper 4366–02.

von Hippel, E (2005). *Democratizing Innovation*. Cambridge: MIT Press.

von Hippel, E and G von Krogh (2003). Open source software and the "private-collective" innovation model: issues for organization science. *Organization Science*, 14(2), 209–223.

von Hippel, E and R Katz (2002). Shifting innovation to users via toolkits. *Management Science*, 48(7), 821–833.

von Hippel, E, S Thomke and M Sonnach (1999). Creating breakthroughs at 3M. *Harvard Business Review*, 77(5), 47–57.

Wasserman, S and K Faust (1994). *Social Network Analysis: Methods and Applications*. Cambridge: University Press.

Wecht, CH (2005). *Frühe aktive Kundenintegration in den Innovationsprozess*. Wien: Alwa & Deil.

Wegner, DM (1987). Transactive memory: a contemporary analysis of the group mind. In *Theories of Group Behavior*, B Mullen and GR Goethals (eds.), pp. 185–208. New York: Springer.

Wegner, DM (1995). A computer network model of human transactive memory. *Social Cognition*, 13(3), 319–339.

Wegner, DM, R Erber and P Raymond (1991). Transactive memory in close relationships. *Journal of Personality and Social Psychology*, 61(6), 923–929.

Weisberg, R (1986). *Creativity: Genius and Other Myths*. New York: Freeman.

Wellman, B and M Gulia (1999). Virtual communities as communities: net surfers don't ride alone. In *Communities in Cyberspace*, MA Smith and P Kollock (eds.), pp. 167–194. London: Routledge.

Wheelwright, SC and KB Clark (1992). *Revolutionizing Product Development: Quantum Leaps in Speed, Efficiency, and Quality*. New York: Free Press.

Whittaker, S, E Issacs and V O'Day (1997). Widening the net: workshop report on the theory and practice of physical and network communities. *SIGCHI Bulletin*, 29(3), 27–30.

Wikström, S (1996). The customer as co-producer. *European Journal of Marketing*, 30(4), 6–19.

Wright, J (2006). *Blog Marketing als neuer Weg zum Kunden: Mit Weblogs die Kunden erreichen, die Marke stärken und den Absatz fördern*. Heidelberg: Redline Wirtschaft.

Zerfaß, A and J Bogosyan (2007). Blogstudie 2007: informationssuche im Internet: Blogs als neues Recherchetool. In http://www.lunapark.de/fileadmin/studien/Uni_Leipzig_blogstudie2007_ergebnisbericht.pdf, [28 March 2008].

Zimbardo, PG (1992). *Psychologie*. Berlin: Springer.

PROACTIVE INVOLVEMENT OF CONSUMERS IN INNOVATION: SELECTING APPROPRIATE TECHNIQUES

KAREN L. JANSSEN and BEN DANKBAAR

Radboud University Nijmegen, The Netherlands

Introduction

Companies in competitive sectors need to deliver innovative products to maintain their market share. Preferably, they should develop not just incremental, but also radical product innovations. Involvement of consumers to support the development of radical product innovations is a subject of debate. Some critics state that consumers do not know what they want in the future (Ulwick, 2005) and cannot formulate those needs (Ciccantelli and Magidson, 1993). Consumers are notoriously lacking foresight, since, according to Hamel and Prahalad (1994), they cannot imagine something that does not exist. Consumers can only make suggestions for improvements of existing products (Veryzer, 1998; Grunert, 2005; Lagrosen, 2005) and involvement has only value for the development of incremental product innovations (Utterback, 1995; Christensen, 1997). Firms can lose their position of industry leadership, if they *listen* too carefully to their customers (Christensen and Bower, 1996). Supporters state that it is very well possible for consumers to support the development of radical product innovations (Leonard-Barton, 1995; Urban *et al.*, 1996; Eliashberg *et al.*, 1997; Wind and Mahajan, 1997; Slater and Narver, 1998). In such situations, consumers have to be stimulated and encouraged to step "out of their box" (Lukas and Ferrell, 2000), and not to be restricted to technological possibilities. Some argue that a high level of consumer involvement is required for the development of radical product innovations (Von Hippel, 2001; Von Hippel and Katz, 2002), and user interaction is critical (Hsieh and Chen, 2005; Lagrosen, 2005). The nature of this type of involvement is proactive: it discovers opportunities for customer value of which the customer is unaware (Kumar *et al.*, 2000; Narver *et al.*, 2004). Although we can also think of situations in which consumer involvement will be of little use or even detrimental, we will follow the lead of those who think that it will be useful in most cases, because "continuing involvement of consumers with developers in an integrated fashion sustains the melding of consumer needs with technical capabilities" (Saguy and Moskowitz, 1999: 70).

The involvement of consumers (by need inputs, concept reviews and product tests) contributes to the superiority of a product (e.g., Cooper and Kleinschmidt, 1993; Chandy and Tellis, 2000; Van Kleef et al., 2005), but more insight into the use of techniques to obtain consumer input is required (Veryzer, 1998). Moreover, the issue of selecting techniques for customer involvement in product development is not a matter of selecting a single technique, but a matter of designing a whole system of techniques linked together in an overall process of consumer involvement in the product development (Kaulio, 1998). To our knowledge, little work has been done about the selection of appropriate techniques that involve consumers in the new product development, covering all phases of the development process *and* specified for types of radicalness.

To improve the utility of consumers' contributions, both R&D and marketing specialists should participate (Hamel and Prahalad, 1994; Veryzer, 1998). Involvement of consumers and interaction across multiple departments, especially between the two departments of marketing and R&D, will provide new opportunities for product development (O'Connor, 1998; Rice et al., 1998; Kahn, 2001). Within this paper, we focus on proactive involvement of consumers by both R&D and marketing specialists of the company.

In the next section, we will first discuss three types of radical product innovation, based on a technology-need matrix, applicable to consumer markets. We will also clarify the development process of radical innovation, for which we distinguish three main phases. In the subsequent section, we offer a discussion on previous categorisation schemes of consumer research. Furthermore, we will explain the requirements for proactive consumer-involvement techniques in six situations of radical product development. Finally, we present the findings of comparative case studies of nine product innovations, where we relate the involvement of consumers to type of radicalness and phases in the development process. We finish this paper with a discussion of the process of selecting the most appropriate techniques, and some concluding remarks.

Radical Product Innovation

Three types of radicalness

Innovations are often analysed in terms of dichotomies: *incremental* and *radical* (e.g., Green et al., 1995; McDermott and O'Connor, 2001), *competence enhancing* and *competence destroying* (Tushman and Anderson, 1986), *continuous* and *discontinuous* (Veryzer, 1998) and *sustainable innovation* and *disruptive innovation* (Christensen, 1997). For the classification of types of innovations, a matrix with two dimensions is often used: a market dimension (existing and new) and a technology dimension (existing and new). But there is more to it. Product innovations can be seen from different perspectives: micro and macro perspective (Garcia and Calantone, 2002) or firm vs. customer perspective (Danneels and Kleinschmidt, 2001).

Garcia and Calantone (2002) argue that a distinction should be made between macro and micro perspectives. From a macro perspective, product innovations introduce industry-wide or market-wide newness, which results from a change in technology and/or market structure in a sector. From a micro perspective, product innovations are new to the firm or new to the consumers. Innovations results from a change in the firm's existing marketing resources or R&D resources. From a marketing viewpoint, product innovations may require new marketplaces to evolve. From a technological viewpoint, the production processes or required equipment are emphasised. For some products, of course, both marketplace and technological factors can change at the same time (Garcia and Calantone, 2002). Basically, these authors distinguish *incremental products*, (occur only at micro level with a minor change in marketing and/*or* technology), *really new products* (include new technologies to existing markets *or* existing technologies to new markets), and *radical innovation* (discontinuities in *both* the existing market structure and the existing technology structure on both macro and micro level).

Danneels and Kleinschmidt (2001) argue that customers and firms perceive innovation differently, and therefore view innovation in different ways: the customer perspective and the firm perspective. From the customer perspective, innovativeness is related to new product attributes, adoption risks and level of change in established behaviour patterns. Products that have a minimal effect on behaviour patterns are incremental product innovations, like a new yoghurt flavour (*coconut yoghurt*). Radical product innovations are disruptive in the sense that they alter prevailing consumer habits and behaviours in a major way (Eliashberg *et al.*, 1997; Markides, 2006), like the one-cup of coffee machines (*Senseo*®). Within the firm perspective, innovativeness is related to environmental familiarity and project-firm fit both in technology and marketing aspects (Danneels and Kleinschmidt, 2001). Newness along one dimension does not imply newness along the other dimension; product innovations can be new to the firm or new to the consumer, or new to both. By this perception, a new technology for a firm does not necessarily have to result in a radical product innovation when it is viewed from the consumer perspective. For instance, a new kind of ice-cream that remains soft and can directly be served from the freezer (Mona Schepijs), has been developed with a new technology for the firm, but the product has minimal effect on the consumer's behaviour.

Urban and Hauser (1993) make use of a technology-need matrix (known to unknown dimensions) to argue that four situations occur in new product development. Also Tidd *et al.* (2001) distinguish four quadrants, each of which raises different issues during development, based on a matrix with maturity of technologies and of markets as dimensions.

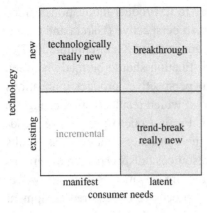

- *breakthrough*: a new product that is developed with a new technology and fulfils a latent consumer need; this type of product innovations has a major effect on behaviour patterns and changes existing technological resources.
- *technologically really new*: a new product that is developed with a new technology and fulfils a manifest consumer need; this type of product innovations changes existing technological resources.
- *trend-break really new*: a new product that is developed with an existing technology and fulfils a latent consumer need; this type of product innovations has a major effect on behaviour patterns (new trend in behaviour).

Fig. 1. Three types of radical product innovations.

Relying on these findings, we propose three types of radical product innovations in consumer markets as visualised in Fig. 1. The technology dimension is divided into "existing" and "new", viewed from the firm perspective at a micro level. The consumer need dimension is divided into "manifest" and "latent", (Narver *et al.*, 2004), viewed from the consumer perspective at a macro level. Manifest needs are needs that can be expressed by people; latent needs cannot (at least not consciously) (Narver *et al.*, 2004).

The development process

In accordance with the literature (e.g., Cooper and Kleinschmidt, 1993; Veryzer, 1998; Tidd and Bodley, 2002; O'Connor, 2005; Vuola and Hameri, 2006), the development process of radical product innovations can roughly be divided into three phases: (1) the discovery phase, (2) the incubation phase and (3) the commercialisation phase. In the *discovery* phase, product opportunities are identified and the readiness of new technologies is established. Opportunities must be generated, recognised and evaluated. In the *incubation* phase, selected product concepts evolve into business propositions and the prototype is designed. Prototypes are to be developed and tested to become new products that are validated and can be launched in the market. Ideas of what the technology platform could enable in the market, and what the market space will ultimately look like will become familiar. In the *commercialisation* phase, the prototype moves out of the R&D department to the operation unit for "scaling up" and the technology is re-defined for a specific application in mass production. The business is ramped up to stand on its own.

The focus of the involvement of consumers will be different over the course of the development process (O'Connor, 1998; Veryzer, 1998). During the *discovery* phase, the focus tends to be on the identification of product opportunities as well as

Table 1. Six situations in the development of radical product innovations.

Type of radicalness	Discovery	Incubation
Technologically really new	Situation 1	Situation 2
Trend-break really new	Situation 3	Situation 4
Breakthrough	Situation 5	Situation 6

on learning about new markets. Consumers are involved from the start to enrich and ensure the relevance of new products (Hise *et al.*, 1989; Gassmann *et al.*, 2006). During the *incubation* phase, the focus is more on the design of the prototype. The underlying technology is further explored and the first try-outs of the product take place. Consumers are involved, for the generation of additional information on customer needs and specifications of the final product. During the *commercialisation* phase, consumers are no longer involved to shape the product, but for clarifying target markets (Veryzer, 1998) and to forecast sales. In this phase, consumer involvement is not essentially different from that in less radical or incremental innovations. We will concentrate on the first two phases and argue that for each type of radicalness and phase in the development process (six situations in Table 1), certain techniques for involving consumers will be most appropriate. For selecting the right technique, it is necessary to characterise the different techniques and analyse the requirements for each situation.

Differentiating and Selecting Techniques

Earlier efforts

Some efforts have been undertaken in the past to classify various techniques for involving consumers in the new product development process. Eliashberg *et al.* (1997) have investigated the role that market research methods can play, in aligning marketplace needs with technological potential. They suggest four methods[1] that may have some value for new products based on familiarity with the product category, relevance of the information, type of information, effect of the knowledge on the response. Urban and Hauser (1993) suggest that the emphasis in development process is different for various situations: for the development of product innovations with unknown technologies, R&D is critical; and when the needs are unknown, marketing research[2] is important. Van Kleef *et al.* (2005) reviewed 10

---

[1] Brainstorming methods, the lead-user technique, information acceleration and methods based on virtual reality.
[2] Lead-users and information acceleration.

techniques[3] in the early stages of the new product development based on information source for need elicitation (stimuli that are used to guide participants in revealing their opinion), task format (the application of the method) and response (the use of the gathered information). Kaulio (1998) reviewed seven different methods[4] for customer involvement in the product development. His results indicate that different methods support the involvement of customers at different phases of the design process (points of interaction between customers and the process) and at different ways (the depth of customer involvement).

These previous categorisation schemes provided starting points for making distinctions in the use of techniques (e.g., in relation to actionability for marketing vs. R&D), however, they do not look systematically at the way consumers are involved in relation to the type of radicalness *and* phase in the development process. Our attempt is to focus on the requirements for each situation and relate consumer-involvement techniques to the types of radicalness and phase in the development process.

Differentiating Characteristics

Thus far, classifications of techniques have been based on differences in the composition of participating groups, the way in which information is acquired, and the way the technique is applied. Starting from these earlier efforts, we have identified four characteristics to distinguish techniques in which R&D and marketing proactively involve consumers to support the development of radical product innovations: *participants*, *stimuli*, *interaction* and *outcome* (*cf.* Fig. 2 and further explanations in Table 2).

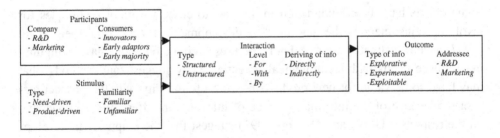

Fig. 2. Differentiating characteristics of techniques that involve consumers proactively.

---

[3]Category appraisal, conjunct analysis, empathic design, focus group, free elicitation, information acceleration, Kelly repertory grid, laddering, lead-user technique, and Zaltman metaphor elicitation technique.
[4]Quality function deployment, user-oriented product development, concept testing, beta testing, consumer idealised design, lead-user method, and participatory ergonomics.

Table 2. Explanation of the differentiating characteristics.

		Explanation
Participants Based on: Urban and Von Hippel (1988), Rogers (2003).	**Company** *R&D*	Tasks of specialists involve shaping new technologies; finding new applications of an existing technology and making new technologies understandable to the consumer.
	Marketing	Tasks of specialists involve shaping consumer needs; elicit future consumer needs, and make sure that the consumer need is being fulfilled by the product.
	Consumers *Innovators*	Consumers in this group are eager to try new things and usually have the ability to understand and apply complex technical knowledge.
	Early adaptors	Consumers in this group tend to be more integrated in real life settings than others, and are social leaders and well educated. These consumers provide advice and information sought by other adopters and they have the reputation to successfully decide on improvements.
	Early majority	Consumers in this group will adopt new ideas just before the average consumer. These consumers think carefully before accepting new ideas and have many informal social contacts.
Stimulus Based on: Van Kleef et al. (2005), Patnaik and Becker (1999), Alba and Hutc-hinson (1987).	**Type** *Need-driven*	With this type of stimuli, consumers are not being exposed to concepts of the products, but problems/verbal communication are the source to elicit information.
	Product-driven	With this type of stimuli, consumers are confronted with (concepts of) products and thereby motivated to provide information.
	Familiarity *Familiar*	The stimuli is recognisable for the consumers, as they have a reference frame for this — the stimulus.
	Unfamiliar	Consumers have no experience with the stimulus, and do not recognise the stimuli by themselves.

Table 2. (*Continued*)

		Explanation
Interaction Based on: Churchill (1995), Van Kleef et al. (2005), Kaulio (1998).	Type *Structured*	This type of interaction refers to gaining response that is highly pre-determined; choices between alternatives can be made. The response is clear to the new product developers.
	Unstructured	This type of interaction refers to an open discussion without pre-set answer categories; consumers can respond in their own words. In-depth and detailed responses have to be analysed by new product developers.
	Level *For*	With this level of interaction, products are designed based on customer research but the customer is not further involved.
	With	With this level of interaction, customer can react upon displays of different concepts and discuss their opinions.
	By	With this level of interaction, customers are actively involved and participate intensively in the product design.
	Deriving of info *Directly*	New product developers derive information directly from the consumers, for example, by means of articulation.
	Indirectly	New product developers derive information indirectly from the consumers, for example, by means of observation.
Outcome Based on: March (1991), Veryzer (1998), Van Kleef et al. (2005), Smith and Tushman (2005), Rochford (1991)	Type of info *Explorative*	This type of information is about understanding what drives consumers' decision processes, and about identifying which factors influence these processes. Information reveals futures that may be quite different from what is presently known.
	Experimental	This type of information is about concrete input for subsequent technical development stages. Trial-and-error, variation, flexibility and playing are characteristics of generating this type of information.

Table 2. (*Continued*)

		Explanation
Addressees		
	R&D	Specialists in this field are more product-oriented and require very concrete information about how to translate consumer-desired product attributes into target values for technical development.
	Marketing	Specialists in this field are consumer need-oriented and require information about life-styles and consumption patterns in the future.

Requirements for consumer involvement in different situations

For various situations in the development of radical product innovations, we argue that a specific combination of differentiating characteristics will be most appropriate. For instance, stimuli can be need-driven (problems are used to elicit information) or product-driven (product concepts are provided to elicit information). In the discovery phase of technologically really new product innovations (new technologies offer new solutions for manifest needs), need-driven stimuli will be more appropriate since existing problem will be connected to new technologies. For each situation, we have developed a configuration of appropriate differentiating characteristics. These specific requirements for consumer-involvement techniques are summarised in Table 3.

For example, the "*ideal consumer session*" to be used in the discovery phase of trend-break really new product innovations has the following requirements. The session is set up mainly by marketing to elicit latent consumer needs, some R&D specialists participate to learn about new applications of the existing technology. Consumers who participate will adopt new ideas just before the average consumer. They think carefully before accepting new ideas and have many informal social contacts. Products (or concepts) are used to elicit information, since they are developed with an existing technology. Therefore, the consumers are somewhat familiar with the stimuli. Participants from the company and consumers work closely to come up with new information. The technology is known and by reflecting on ideas or concepts, information on latent consumers' needs might be generated. Consumers are free to give any type of information, since new direction must be found. The information has to be derived indirectly from the consumers, as consumers in this situation are not fully aware of their needs. New product developers generate information on product opportunities and new markets. Results are to be used mainly by marketing to secure a market (fulfillment of latent needs).

Table 3. Requirements for consumer involvement.

	Technologically really new		Trend-break really new		Breakthrough	
	Discovery	Incubation	Discovery	Incubation	Discovery	Incubation
Participants						
Company	Mainly R&D	Mainly R&D	Mainly marketing	Mainly marketing	Both R&D and marketing	Both R&D and marketing
Consumers	Early adopters	Early majority	Early majority	Early majority	Innovators	Innovators
Stimulus						
Type	Need-driven	Product-driven	Product-driven	Product-driven	Need-driven	Product-driven
Familiarity	Familiar	Unfamiliar	Familiar	Familiar	Unfamiliar	Unfamiliar
Interaction						
Type	Unstructured	Structured	Unstructured	Structured	Unstructured	Unstructured
Level	"For"	"For"	"With"	"With"	"By"	"By"
Deriving of info	Directly	Directly and/or indirectly	Indirectly	Directly and/or indirectly	Indirectly	Directly and/or indirectly
Outcome						
Type of info	Explorative	Experimental	Explorative	Experimental	Explorative	Experimental
Addressees	Mainly R&D	Mainly marketing	Mainly marketing	Mainly R&D	Both R&D and marketing	Both R&D and marketing

Appropriateness of 20 techniques

Each technique that involves consumers had its own characteristics. We have reviewed the literature and established the differentiating characteristics of 20 techniques that involve consumers proactively.[5] These 20 techniques are briefly described in the Appendix. The differentiating characteristics allow us to combine the techniques with the three types of radicalness and phase in the development process. We positioned the techniques against each situation and calculated the number of fits between the requirements of the situation and the characteristics of the technique (precise calculations are available from the authors). Table 4 lists the most appropriate techniques for each situation.

Table 4. Most appropriate consumer-involvement techniques for each situation.

Technologically really new	
Discovery-phase	*Incubation-phase*
Lateral thinking synectics	Category appraisal
	Conjoint analysis
	Information acceleration

Breakthrough	
Discovery-phase	*Incubation-phase*
Applied ethnography	Applied ethnography
Crowd sourcing (develop)	Crowd sourcing (design)
Toolkit for innovation	Lead-user method (need)
Visioning/back casting	Toolkit for innovation

Trend-break really new	
Discovery-phase	*Incubation-phase*
Applied ethnography	Applied ethnography
Consumer idealised design	Probe and learning
Empathic design	
Innovation templates	

---

[5] Several techniques have been excluded after reviewing, because the central element "interaction between company participants and consumers" is very low, or because they only result in information on incremental improvements or new ideas. Among them are focus group, brainstorming, quality function deployment, laddering, Kelly Grid analysis, beta testing, nominal group method, free elicitation and sensory methods.

Case Studies

Methodology

To gain further insights into the differentiating characteristics of consumer involvement in different phases of development and for different types of radicalness, we used a historical comparative case-study approach (Yin, 2003). By collecting detailed information about the role of consumer involvement in various cases of radical product development, we hoped to find confirmation for the distinctions made above as well as additional information about the characteristics of techniques for the proactive involvement of consumers. We analysed the development of nine radical product innovations in the Dutch food sector (food product, year of introduction: Becel Proactive 1998; Breaker 2001; Senseo 2001; Fruit2Day 2004; MonaSchepijs 2004; KnorrVie 2005; Valess 2005; Fridéale 2006; Optimel Control 2006). Because of confidentiality, no direct references to the products are made. The nine product innovations were selected on the basis of a preliminary investigation with six experts in the food domain: 20 products that have been introduced in Dutch supermarkets since the year 1998 were classified on the type of radicalness. The selection out of these 20 was based on agreements in the opinion of the experts. We selected three products for each type of radicalness for a total of nine and analysed the development of these products.

It was of crucial importance to understand the circumstances of the sessions that involved consumers. Therefore, we incorporate the point of view from marketing as well as from R&D. In-depth-interviews were held with the responsible R&D manager and marketing manager at the time of development, or persons with comparable responsibilities. Data were collected on the involvement of consumers from discovery of the idea, to development of the prototype and commercialisation of the product. Special attention was paid to understand the characteristics of the way consumers were involved to support the development of the final product.

Findings

Technologically really new product innovations

The development of technologically really new product innovations started because of the urge to extend a product category, driven by a technological push. To find a market for the technology, developers anticipated on general trends in the consumer market, like *"healthy food"* or *"more convenience"*. An important reason to involve consumers in the development of this type of radicalness was the use of consumers' perspective by constructing the technology. One respondent argued: *"The point of difference in the existing product category has to be found, but also, this point needs to be relevant"*. Another respondent explained *"During the category research, it was not the first thing that consumers indicated, but when we asked for unpleasant aspects in consuming the product, people agreed".*

The company that developed TRN1 conducted brainstorm sessions with employees on the technological feasibility since they were convinced of the potential success of the concept. They talked about this specific concept through with an advertising agency, as representatives of the consumers, to generate information on expectations of the success of the concept. The company believed so strongly in the concept that they decided to continue the development without actual consumer research. The company that developed TRN2, first developed a prototype of the product to explore the technological feasibility before involving consumers. The company that developed TRN3 followed a different path. An explorative research of the product category, that was initiated by marketing, in which participants discuss several existing products, shed light on the unpleasant aspects of the category. Approximately 20 consumers participated: consumers who liked to participate in sessions and meet the requirements of the supposed target group, and a group with the direct opposite characteristics of the supposed target group. The consumers grouped multiple existing products of the category under study, discussed hindering aspects and explored possible improvements. Marketing was addressed to formulate the statement of the concept. Comparing our definition of the discovery phase, this test was conducted in a more pre-development phase, since more than one existing products were evaluated to find a new concept for the product category.

The development of TRN1 as well as TRN2, since no consumer research was conducted in the discovery phase, was supported with a combined concept-product test. Consumers were involved in one session that involved both concept review and product tests. Participating consumers, about 120–150, were selected on expectations of the target group. Other requirement was that they often buy or consume products in the category under study. The session of TRN1 started with a product test. Multiple existing products and the new product were tasted and judged. The session ended with a discussion on the new product. Marketing learned about the formulation of the concept and R&D could make some improvements of the product. The consumer session of TRN2 started with general questions on consumption patterns, followed with a discussion about unpleasant aspects of the products. The concept of the product was presented and initial reactions were asked. There were no pre-set answer categories. Very indicative information on *"what's in it for me and you"* rather than *"do you like it"* was generated. Marketing was addressed to formulate the concept and learn on the communication strategy. In the next stage of the session, the product was presented and consumers could taste and evaluate the product. The interaction with consumers became more structured; consumers had to score several aspects of the products. At this point, R&D was addressed with information to optimise the product. Improving the taste was one of the main issues to generate information on. Consumer involvement in the incubation phase of the development of TRN3 was similar as described above. The first stage of the session was an open group discussion on concept statement: *"We asked the consumers if they could think along with us about the new concept for the category. The consumers were able to provide a lot of inputs. The concept was shaped step by step."* Marketing learned about the communication strategy. R&D observed to learn about the new category. In a second session, a structured interview with individuals shaped the product further. R&D was mainly addressed with information to optimise taste.

Table 5 summarises the differentiating characteristics of proactive consumer involvement in the development of the technologically really new product innovations under study. The concept test and the product test of TRN2 were conducted in two stages, with different stimuli and interaction. Therefore, this concept–product test has been split up into two columns. The group test and individual test of TRN3 have also been split up in two columns. Although the respondent explained that both tests were applied in the incubation stage, and the pre-development test in the discovery phase, we summarised the result differently: the context of the group test exercise was more corresponding to the discovery phase. The context of the individual test corresponded more to the incubation phase.

Trend-break really new product innovations

The development of trend-break really new product innovations started in all cases with the aim to extend a brand by introducing new products. Product developers proposed new concepts within the brand to refresh the image. They translated consumer response into statements that a consumer could identify himself/herself with. The interpretive skills of new product developers are important to envision future lifestyles of the consumers, as one respondent argued *"Although the product did not exist, the orientation of development is a kind of reactive. It sounds very conservative, but we never innovated because consumers told us what they wanted. We only get answers to the questions you ask, but not to the questions you do not ask"*.

The creation of the concept of TBRN1, TBRN2 and TBRN3 was due to internal discussion among colleagues, no consumers were involved to discover ideas. Concepts where built in multidisciplinary brainstorm sessions with employees and some external contacts, which were closely associated to the company, were consulted. Both internal and external reactions were ultimately so positive that the concept was shaped without the involvement of consumers. However, in a later stadium of the discovery phase consumers were involved to evaluate the concept. The company that developed TBRN1 provided a maximum of 15 consumers with the statement of the concept on boards, and formed the story around the concept: *"The pattern of thoughts of the concept had to fit with the lifestyle of the consumers. The concept*

Table 5. Consumer involvement in the development of TRN1, TRN2 and TRN3.

	TRN1	TRN2		TRN3		
	Incubation	Incubation		Discovery	Incubation	
	Concept test *(discovery)*	Product test *(incubation)*		Pre-test	Group test *(discovery)*	Individual test *(incubation)*
Participants						
Company	Both marketing and R&D	Both marketing and R&D		Marketing	Both marketing and R&D	Both marketing and R&D
Consumers	n/a	n/a		n/a	n/a	n/a
Stimulus						
Type	Need-driven/ product-driven	Product-driven		Product-driven	Need-driven/ product-driven	Need-driven/ product-driven
Familiarity	Familiar	Unfamiliar		Familiar	Familiar (need)/ unfamiliar (product)	Familiar (need)/ unfamiliar (product)
Interaction						
Type	Unstructured	Structured		Unstructured	Unstructured	Structured
Level	With	With		With	With	With
Deriving of info	Directly	Directly		Directly	Directly	Directly
Outcome						
Type of info	Explorative	Experimental		Explorative	Explorative	Experimental
Addressees	Marketing	Both marketing and R&D		Marketing	Both marketing and R&D	Both marketing and R&D

was somewhat familiar, but intractable, thus a good explanation was needed." In an open discussion, the credibility of the claim, the minimal requirements, the intention to buy and the willingness to change consumption patterns were discussed. The company found it surprising that a large group of consumers did not have the required power of imagination as expected: "*The claim of the concept seemed so logical that it was strange to notice that people did not follow the recommendations of the government*". Consumers were involved in the discovery phase of TBRN2. The concept was solely tested on paper. Several different statements of the concept were discussed with a group of 15 consumers: "*Beforehand, we could imagine a certain amount of directions of the concept. The story statement could be 'very taste-dominant and the other aspects reason-to-believe' or the story was really about 'the functional aspect and taste as side aspect'. There were a few buttons we could turn.*" This session was repeated three times in three countries. The interaction was very unstructured and resulted sometimes in unexpected findings: "*We were surprised to see that consumers did not believe the claim of the product, or did not acknowledge the problem they ought to have.*" Marketing had to filter the concept statement out of the reactions of the consumers. To support the development of TBRN3, the concept was first explained to a small group of consumers. Then, the consumers worked with the concept and mapped their needs and the fit with the concept was made. An open discussion followed and consumers could express their approval or disapproval. Based on the outcome of the session, TBRN3 was refined to consumer needs.

In the incubation phase, a large group of consumers were involved to support the development by judging the product on details. A panel, about 100–200 people of the expected target group, evaluated one or more prototypes on taste, texture and other aspects, like packaging and size. At the beginning of the session, consumers could not recognise the product, because it did not yet exist. Only in consumer sessions of TBR3, some consumers could recognise the product, since they already participated in earlier sessions: "*A number of consumers did come back a few times. The first time, they were unfamiliar with the concept and the product, but at later sessions, the products became familiar. We even asked them what kind of improvements they observed and what they thought of it.*" Both marketing and R&D were involved in the design of the sessions, as pointed out by one respondent (TBRN3): "*The consumers have to see the concept in its context, and therefore, the interaction starts with an open discussion about the statement of the concept and ends with closed questions on price, consumption and product specification.*" R&D was addressed to refine the product and marketing was addressed to communicate the concept. A remarkable issue that came up during the consumer session of TBRN1 was the moment of consumption of the product: "*The consumers pointed out clearly that they wanted to consume the product in the out-of-home segment. They wanted to be able to purchase the product at school, at train stations and petrol stations. We* had to invest in other supply chains besides the supermarkets." Similar findings emerged during consumer sessions in development of TBRN2: *"Consumers asked why we did not use another packaging of the concept, but the packaging was the fundamental statement of the concept! We do not blindly follow the consumer. We learned to refine the communication."*

The differentiating characteristics of the proactive involvement of consumers in the case of trend-break really new product innovations are summarised in Table 6.

Breakthrough product innovations

Findings suggest that breakthrough product innovations are also developed from research within a specific product category. Important reasons for proactively involving consumers were to develop new reference frames, give face to the concept idea and test the credibility of the claim of the concept. The actual product, "the technical state" was more or less fixed and marketing wished to learn about communicating the concept to the consumer. A remarkable comment: *"It is very difficult to ask consumers questions about a product that is unfamiliar to them. The outcome is very much dependable of the visionary skills of the marketeers. The interpretation of the answers matters."* Some respondents shared the opinion that in many cases the interpretation of the outcome is not properly done; missing opportunities are the result.

In the discovery phase of breakthrough product innovation, all companies did a concept tests with consumers. However, the concepts were presented in different ways. The concept of B1 was developed out of preliminary research within a specific product category: consumption patterns of products in this category were analysed in several countries in the world. In a group session with consumers, marketing discussed the product category with consumers, and then showed a mock-up and discussed the specific idea for the new product. The mock-up was used to demonstrate that B1 was different than existing products in that category. A small group of consumers, about six, was asked to make mood boards, a kind of collage, of different concepts (existing ones and the new one) and discussed the results. This session was repeated four times with new groups of consumers. They also had to make a value map. Marketing looked for consistency in the concept, the product, and later packaging, to position the product. This consistency was very important: *"We have to interpret what fits best with the concepts instead of what consumers like at the moment. What they like at the moment is reactive, not proactive. The consistency in the statement of the concept matters for the right positioning of the product."* The concept test of B2 was conducted with the same purpose. Marketing had developed eight statements of the concepts and wanted to find out which statement consumers met with approval. For each single statement, a session was organised with 10 consumers of the same target group (different for each session).

Table 6. Consumer involvement in the development of TBRN1, TBRN2 and TBRN3.

Trend-break really new

	TBRN1		TBRN2		TBRN3	
	Discovery	Incubation	Discovery	Incubation	Discovery	Incubation
Participants						
Company	Marketing	Both R&D and marketing	Marketing	Both R&D and marketing	Marketing	Both R&D and marketing
Consumers	n/a	n/a	n/a	n/a	n/a	n/a
Stimulus						
Type	Need-driven	Product-driven	Need-driven	Need-driven/ product-driven	Need-driven	Need-driven/ product-driven
Familiarity	Unfamiliar	Unfamiliar	Unfamiliar	Unfamiliar	Unfamiliar	Unfamiliar
Interaction						
Type	Unstructured	Structured/ unstructured	Unstructured	Structured	Structured	Structured/ unstructured
Level	With	For	With	For	With	With
Deriving of info	Indirectly	Directly	Indirectly	Directly	Indirectly	Directly
Outcome						
Type of info	Explorative	Experimental	Explorative	Experimental	Explorative	Experimental
Addressees	Marketing	Both R&D and marketing	Marketing	Both R&D and marketing	Marketing	Both R&D and marketing

Each statement was expressed as a range of questions on a card, from very general to really specific questions. Both marketing and R&D observed the discussion and were able to interact by means of a moderator. Information on interest in the product was generated, even as information on the range consumers would be willing to pay for the product. The company learned that the product will be accepted or not be accepted depending on how you communicate the concept: *"As soon as you fall into the 'authority of science', thus when you ask a scientist to present the concept, and let him tell that 'science has approved that . . .' you will lose the consumers. They will brush it aside, they will not believe it. We had to find another way to communicate our findings, and we did! . . . When we saw a consumer reaction 'hey, that concerns me, that is what I need', we knew what the formulation of concept statement should be. We looked for the insights of the consumers to accentuate the credibility of the claim."* The concept test of B3 was done with an explanation on paper. On the computer, consumers had to score a range of aspects of the concepts. The aspects with the highest scores were linked and applied in the final concept. This concept was discussed with marketing and R&D and information on the communication of the concept was derived.

The central element of involving consumers in the incubation phase is the at-home test. Consumers, about 100–150, have to use the product in daily circumstances and record their consumption behaviour. Open questions, as well as closed questionnaires, have to provide fixed data on product specifications, use-issues and taste. Marketing is addressed with information on the communication of the product. A solid basis for launching the product must be found: *"We have to find a benchmark: the target group will be . . . , this number of consumer will buy the product, and the price has to be . . . . We are looking for a frame of reference."* In some, R&D was addressed to improve the prototypes.

The differentiating characteristics of the proactive involvement of consumers in the case of breakthrough product innovations are summarised in Table 7.

Reactive research in the commercialisation phase

In the final phase, no proactive consumer research was conducted. Information to validate product performance and estimate sales figures was conducted with reactive questionnaires on a large scale. Closed questionnaires provided detailed information about uniqueness, intention to buy, price ranges and the aimed target group. Some of these aspects have already been considered in previous research, for instance price: *"We test the price range in the early stage. If the consumer is willing to pay more, we can use better and more expensive ingredients"*, but these aspects need to be tested again with a large group of consumers. The final marketing package is checked by marketing to formulate the distinguishing features of the concept.

Table 7. Consumer involvement in the development of B1, B2 and B3.

	Breakthrough					
	B1		B2		B3	
	Discovery	Incubation	Discovery	Incubation	Discovery	Incubation
Participants						
Company	Marketing	Marketing	Both R&D and marketing	Marketing	Both R&D and marketing	Marketing
Consumers	n/a	n/a	n/a	n/a	n/a	n/a
Stimulus						
Type	Need-driven/product-driven	Product-driven	Need-driven	Product-driven	Need-driven	Product-driven
Familiarity	Unfamiliar	Familiar	Unfamiliar	Familiar	Unfamiliar	Familiar
Interaction						
Type	Unstructured	Structured/unstructured	Unstructured	Structured/unstructured	Unstructured	Structured/unstructured
Level	By	For	By	For	By	For
Deriving of info	Indirectly	Directly	Directly/indirectly	Directly	Directly	Directly
Outcome						
Type of info	Explorative	Experimental	Explorative	Experimental	Explorative	Experimental
Addressees	Marketing	Both R&D and marketing	Marketing	Marketing	Both R&D and marketing	Both R&D and marketing

Noticeable comment of the development of TRN2 was that up to three times after introduction, the statement of the concept was adjusted: *"The point of difference was extremely difficult to find since people had to be convinced to change attitude and behaviour in consumption patterns. Incremental innovations in the new product category are responsible for the success of this product."* Remarkably, developers of TRN3 did not follow-up the product category which is probably the reason for the low success rate of the product.

Selection of consumers

For each situation, we distinguished a specific type of consumers that should be involved. The selection was based on Rogers' adoption curve that classified consumers into categories (innovators, early adopters and early majority) based upon how long it takes them to adopt a new idea. We argued that consumers differ in the ability to which they want to make an effort to understand manifest and latent needs and look for solutions, and therefore, that companies needed to select participating consumers on these abilities. Findings indicate that companies do not use of this distinction. Companies just select consumers on demographic characteristics. New product developers envision the target group, and involve consumers with similar characteristics. In some cases, they also select consumers directly opposite to the target group.

Discussion

Technologically really new product innovations

Results show that during the discovery phase, consumer research is not commonly used as starting point for development of technologically really new product innovations. We argued that intensive interaction with consumers might help R&D to develop solutions for manifest needs, and expected that R&D could use these manifest needs of consumers as a starting point for the development of technologically really new product innovations. However, we did not find any consumer-involvement in the discovery-phase that started like this. Product-driven stimuli were still needed to identify opportunities for new technological development, since consumers could not identify the potentials of a new technology. When the development started with a technological invention of R&D, and the confidence of marketing in the application was great, no consumer research was considered to be required to check the findings. However, since the technology was new, and aimed at an existing market, marketing wished to learn about how to communicate the concept. This was usually done in the incubation phase by comparing prototypes of the product with existing products in the same category. It was considered necessary to use existing products as stimulus, since consumers call for a reference frame. If the development of a technologically really new product innovation started with the request of marketing to expand a category, consumer research could contribute to shed light on the unpleasant aspects in an existing product category. Consumers ask for familiar products to express manifest needs. The level of interaction between R&D, marketing and the consumers was more intense than expected. Although the needs were manifest, and consumers were familiar with them, a brief consultation of consumers was not found to be adequate. New product developers had to explain the new product (concept) to the consumers and discuss it in comparison with existing products. The type and deriving of information were as expected: consumers were able to give direct answers, because they had a reference frame, and the interaction was unstructured, since no fixed answer categories existed. The outcome was also as expected: the type of information was explorative, since new product developers searched for a proper communication strategy and the right product specifications. Both R&D and marketing were addressed to learn about communication and product specifications.

Consumer research appeared to be particularly relevant in the incubation phase. The characteristics of the consumer-involvement sessions of the cases under study corresponded with our "ideal consumer session" of this situation. However, some small comments are in place. R&D and marketing equally participated in the session. Marketing wished to learn about the communication strategy and R&D about the optimisation of the product. Contrary to what we expected, a brief consultation of consumers was in this phase not enough, because the new product was still too unfamiliar to the consumer. The way of involvement will always be "with", since new technologies have to be explained and discussed. Remarkable were the consumer sessions after commercialisation of the products of this type of radicalness. Here, consumer-involvement sessions are considered to be reactive. Nevertheless, they gave much insight into the missing aspects of the product innovation. On the basis of these insights, new product developers were able to improve the concept step-by-step. The value of technologically really new product innovations may also depend on the incremental improvements after commercialisation.

Trend-break really new product innovations

Findings show that during the discovery phase of trend-break really new product innovations, consumers are often involved in the last part of this phase because of confidentiality issues. New product developers are too afraid that consumers, or other participants, might steal the idea and sell it to the competition. Ideas are kept secret, until enough efforts are made for a product launch. One of the most important reasons to involve consumers is convincing higher management of the potential of the concept. Therefore, in the discovery phase, consumer involvement focuses on the commercial potential of the concept. The concept is evaluated to learn about the communication strategy and specifications of the actual product. This explains the differences in stimuli as starting points to elicit information on consumer needs. We expected that consumers could and should be involved to discover latent needs by observing and discussing the consumption of existing products (familiar product-driven stimulus) but the case studies indicated an unfamiliar need-driven stimulus. Interaction with marketing was expected to bring latent consumer needs to the surface. We did not find any support for this, within the products of this type of radicalness under study. Although, we see similarities with the characteristics of the consumer involvement in the development of TRN3, we also observed a big difference in development: the development of TRN3 required a new technology to solve a problem. Consumers were able to express the problem, what they felt to be important within this product category. Yet, we still argue that consumers should be involved for the generation of information on latent consumer needs by discussing existing products. By observing consumption behaviour, new product developers might discover opportunities they missed in discussions among employees. The company should not be too afraid to share insights with the consumers, since both the parties could benefit and confidentially agreement can be made.

In the incubation phase, the characteristics of consumer involvement show many similarities among the three products under study. Companies used more stimuli than expected. They used both a need-driven stimulus (to explain the concept) and a product-driven stimulus (as frame of reference to the prototype). The prototype was unfamiliar to the consumers. We expected that the product-driven stimulus would show a large resemblance with existing products, and therefore would be recognised by consumers. However, because of the totally new application of the existing technology and a new way of consumption, need-driven stimuli appeared to be needed as well. Consumers need a combination of a need-driven stimulus (explanation) and product-driven stimuli (the unfamiliar product) in this situation, and adjustments to the ideal consumer-involvement session might be required. Remarkable outcome of the consumer involvement in this phase of the development were information about the moment of consumption. The involvement sessions gave much insight into the consumption behaviour for the product, and marketing was able to specify the launch strategy on the basis of these insights. Both marketing and R&D were addressed. The value of the product innovations in this category may also depend to a great extent on the communication of benefits of the product, for instance health claims.

Breakthrough product innovations

We found most similarities with the requirement for consumer involvement in breakthrough product innovations. Consumer involvement in the development of products under study was nearly the same, as the requirements we developed based on theory. However, consumers were involved at a later moment in the development, because of the reasons of confidentiality. Concept tests were initiated by marketing. Consumers were not involved to find out about new ideas. The main reason to involve consumers was to learn about the proper way to communicate, the technological and functional aspects of the product. The statement of the concept should connect to consumers' perceptions of added value. For instance, a company that develops functional ingredients could communicate the same product to "*prevent*" or "*cure*" something. Identification of the target group is done to specify the communication strategy. For this type of radicalness, no frame of reference exists.

The main difference to the requirements of consumer involvement in the incubation phase, of the development of breakthrough product innovations, appeared to be the familiarity of the stimuli. We expected that the prototype would still be unfamiliar to the consumer, since both technology and consumer need were new. However, the concept was first explained to the consumer, and then the consumer used the product in their daily life for some time. The at-home test was the actual consumer-involvement session to collect information on approval or disapproval of the concept. At this point, the stimulus is familiar to the consumer. But before, the consumer was unfamiliar with the stimulus. Findings indicate that the type of interaction was "for", since at the moment of the at-home test; consumers could not participate in developing the actual product. The company used the outcome to improve the product; therefore, we think that "for" does not give an accurate representation.

Conclusion

The purpose of this paper was to gain insights into proactive consumer involvement, in various situations, in the development of radical product innovations, covering the two main phases of the development process, and specified for three types of radicalness. Based on the literature, we have developed requirements for the involvement of consumers in six situations (Table 3). We positioned 20 techniques from the literature, against each situation and determined the most appropriate techniques for each situation (Table 4). Based on our theoretical study, we conclude that in the discovery phase of *technologically really new* product innovations, creative thinking techniques are appropriate, since they open new ways to solve existing problems. In the incubation phase, techniques to set a hierarchy in product specifications are appropriate since they contribute to the design of product specifications.

In the discovery phase of *trend-break really new* product innovations, observation techniques are appropriate since latent consumer needs can emerge to the surface. Consumers are observed dealing with problems that they were not able to express by themselves. For gaining insights into new ideas, techniques that combine parts of different products are also suitable. In the incubation phase, techniques that specify designs, in one case by observation and in the other case by experimenting seem to be appropriate. For the development of *breakthrough* product innovations, a variety of techniques appear to be appropriate. This is probably due to the fact that both the information on applications of new technologies and information on latent consumer needs are required. In the discovery phase, techniques that depend on the "visionary skills" of consumers are most appropriate. In the incubation phase, the most appropriate techniques depend on the technological insight of a progressive group of consumers or experiment with several versions of product concepts.

Our theoretical study has shown that for each type of radicalness and phase in the development process-specific techniques for proactive involvement of consumers will be appropriate. By using a comparative case-study approach, we evaluated the requirements for the involvement of consumers for each situation. The findings of our comparative case studies in the Dutch food industry, provide evidence that companies proactively involve consumers to support the development of radical product innovations. Moreover, the results indicate that information that is generated by the involvement of consumers is one of the most important factors giving direction to the development of radical product innovations. However, new product developers tend to involve consumers to specify the product (concept) and to determine the launch strategies, rather than involve consumers as a starting point for development. We found support for our proposition that the involvement of consumers depends on phase in the development and type of radicalness. Based on our findings and the discussion, we argue that new product developers should consider the techniques for the involvement of consumers more carefully, and give more thought to the value of selecting the most appropriate technique.

Acknowledgements

The authors gratefully acknowledge fruitful discussions with the colleagues of the MICORD (*Managing Innovation, Outsourcing and Collaboration of Research & Development*) research program, and particularly the comments by Jan de Wit and Geert Vissers. The authors also thank the reviewers of this journal for the comments.

Appendix

Short description of techniques that proactively involve consumers

Applied ethnography	Company participants spend significant periods of time with a consumer group, such as teenagers, retail shoppers, mobile phone users or others, and make detailed observations of their practices to understand their way of life.
Category appraisal[6]	Consumers evaluate a set of competing new concepts by making a visual representation of position that those concepts hold in their mind; the key elements of the market structure as perceived or preferred by consumers are exposed.
Conjoint analysis	Consumers express and rank their preferences toward experimentally varied product concepts that are described by several attributes at several levels; the underlying purchase motivations and trade-offs are exposed.
Consumer idealised design	Potential consumers actually develop an unconstrained design of their ideal product or service; they are not concerned with the feasibility of the design, only with its desirability, and are encouraged to specify ways for revision.
Crowd sourcing design	Consumers are integrated into the design process by using information technology; they test really innovative products for which little consumer experience exists and market research is unclear.
Crowd sourcing development	Consumers are integrated into the development process by using information technology; they develop products for relatively small markets that are further developed by new product developers into production-grade specifications.
Empathic design	Consumers are observed in their own environment, on a specific subject; company participants spend time with them using or consuming products, and grow empathy for the problems that consumers experience.
Information acceleration	Company participants presented the future product and usage scenarios to the consumer in a virtual environment; they constantly provide new information to learn how consumers deal with the product and observe consumers' responses.
Innovation templates	Consumers are provided with components and attributes of products, to make new compositions; templates are used to systematically change the product from its earliest composition to a new product.
Lateral thinking	Participants are trained to change their way of thinking and approach problems in a novel way. By looking at things in different ways, new trains of thought are set off, and a sort chain reaction follows.
Lead-users method (need)	Company participants interact in sessions with lead users, who are seen as representatives of the target-market because they experience today what other consumers will experience months or years later; problems are identified.
Lead-users method (product)	Company participants interact in sessions with lead users, who are seen as representatives of the target-market because they experience today what other consumers will experience months or years later; solutions are identified.

---

[6]Category appraisal is also referred to as internal and/or external preference analysis.

	(Continued)
Probe and learning	A quasi-experimental design to achieve a proper product concept: introducing an early version of the concept to group of consumers, learning from the reactions, modifying the product and approach the consumers and trying again.
Synectics	A creative session in which an original problem is extended into a much wider problem or analogy; for this alternative problem, solutions are generated and later, the solutions are transformed back into solutions for the original problem.
Toolkit for innovation	Consumers are provided with a "user toolkit" that helps them create a product that they want; company participants abandon to find out about consumers' "sticky" need-related information, since consumers carry out the design.
Visioning/ Backcasting	Customers define future events so that the events' occurrences can be interpreted. They mainly focus on "what could be", but not much is known about that yet (visioning — way to the future/back casting — start with the end state).
ZMET[7] (need)	Consumers create collages, based on their feelings and experiences related to a specific problem; participants discuss the images selected and their associated experiences; preferences and specifications are represented in a mental model.
ZMET (product)	Consumers create collages, based on their feelings and experiences related a specific product; participants discuss the images selected and their associated experiences; preferences and specifications are represented in a mental model.

References

Alba, JW and WJ Hutchinson (1987). Dimensions of consumer expertise. *The Journal of Consumer Research*, 13(4), 411–454.

Chandy, RK and GJ Tellis (2000). The incumbent's curse? Incumbency, size, and radical product innovation. *Journal of Marketing*, 64, 1–17.

Christensen, CM (1997). *The Innovator's Dilemma: When New Technologies Cause Great Firms to Fail*. Boston: Harvard Business School Press.

Christensen, CM and JL Bower (1996). Customer power, strategic investment, and the failure of leading firms. *Strategic Management Journal*, 17(3), 197–218.

Churchill, GA (1995). *Marketing Research: Methodological Foundations*. Chicago: The Dryden Press.

Ciccantelli, S and J Magidson (1993). From experience: consumer idealized design, involving consumers in the product development process. *Journal of Product Innovation Management*, 10, 341–347.

---

[7]Zaltman metaphor elicitation technique.

Cooper, RG and EL Kleinschmidt (1993). Screening new products for potential winners. *Long Range Planning*, 26(6), 74–81.

Danneels, E and EJ Kleinschmidt (2001). Product innovativeness from firm's perspective: its dimensions and their relation with project selection and performance. *Journal of Product Innovation Management*, 18, 357–373.

Eliashberg, J et al. (1997). Minimizing technological oversights: a marketing research perspective. *Technological Innovation: Oversights and Foresights*, R Garud, PR Nayyar and ZB Shapira (eds.), pp. 214–230. USA: Cambridge University Press.

Garcia, R and R Calantone (2002). A critical look at technological innovation typology and innovativeness terminology: a literature review. *Journal of Product Innovation Management*, 19, 110–132.

Gassmann, O et al. (2006). Extreme customer innovation in the front-end: learning from a new software paradigm. *International Journal of Technology Management*, 33(1), 46–66.

Green, S et al. (1995). Assessing a multidimensional measure of radical technological innovation. *IEEE Transactions on Engineering Management*, 42(3), 203–214.

Grunert, KG (2005). Consumer behaviour with regard to food innovations: quality perception and decision-making. *Innovation in Agri-Food Systems*, WMF Jongen and MTG Meulenberg (eds.), pp. 57–85. Wageningen: Wageningen University Press.

Hamel, G and CK Prahalad (1994). *Competing for the Future*. Boston, MA: Harvard Business School Press.

Hise, RT et al. (1989). The effect of product design activities on commercial success levels of new industrial products. *Journal of Product Innovation Management*, 6, 43–50.

Hsieh, LF and SK Chen (2005). Incorporating voice of the consumer: does it really work? *Industrial Management and Data Systems*, 105(6), 769–785.

Kahn, KB (2001). Market orientation, interdepartmental integration, and product development performance. *Journal of Product Innovation Management*, 18(5), 314–323.

Kaulio, MA (1998). Customer, consumer and user involvement in product development: a framework and a review of selected methods. *Total Quality Management*, 9(1), 141–149.

Kumar, N et al. (2000). From market driven to market driving. *European Management Journal*, 18(2), 129–142.

Lagrosen, S (2005). Harnessing the creative potential among users. *European Journal of Innovation Management*, 8(4), 424–436.

Leonard-Barton, D (1995). *Wellsprings of Knowledge*. Boston, Massachusetts: Harvard Business School.

Lukas, BA and OC Ferrell (2000). The effect of market orientation on product innovation. *Journal of the Academy of Marketing Science*, 28(2), 239–247.

March, JG (1991). Exploration and exploitation in organizational learning. *Organizational Science*, 2(1), 71–87.

Markides, C (2006). Disruptive innovation: in need of better theory. *Journal of Product Innovation Management*, 23, 19–25.

McDermott, CM and GC O'Connor (2001). Managing radical innovation: an overview of emergent strategy issues. *Journal of Product Innovation Management*, 19, 424–438.

Narver, JC et al. (2004). Responsive and proactive market orientation and new product success. *Journal of Product Innovation Management*, 21, 334–347.

O'Connor, GC (1998). Market learning and radical innovation: a cross case comparison of eight radical innovation projects. *Journal of Product Innovation Management*, 15, 151–166.

O'Connor, GC (2005). Open, radical innovation: toward an integrated model in large established firms. *Open Innovation: Researching a New Paradigm*, H Chesbrough, W Vanhaverbeke and J West (eds.), pp. 62–81. Oxford: Oxford University Press.

Patnaik, D and R Becker (1999). Needfinding: the why and how of uncovering people's needs. *Design Management Review*, 10(2), 37–43.

Rice, MP et al. (1998). Managing discontinuous innovation. *Research Technology Management*, 41(3), 52–58.

Rochford, L (1991). Generating and screening new product ideas. *Industrial Marketing Management*, 20, 287–296.

Rogers, EM (2003). *Diffusion of Innovations*. New York: The Free Press.

Saguy, IS and HR Moskowitz (1999). Integrating the consumer into new product development. *Food Technology*, 53(8), 68–73.

Slater, SF and JC Narver (1998). Customer-led and market-orientated: let's not confuse the two. *Strategic Management Journal*, 19, 1001–1006.

Smith, WK and ML Tushman (2005). Managing strategic contradictions: a top management model for managing innovation streams. *Organizational Science*, 16(5), 522–536.

Tidd, J et al. (2001). *Managing Innovation: Integrating Technological, Market and Organizational Change*. Chichester: John Wiley & Sons.

Tidd, J and K Bodley (2002). The influence of project novelty on the new product development process. *R&D Management*, 32(2), 127–138.

Tushman, ML and P Anderson (1986). Technological discontinuities and organizational environments. *Administrative Science Quarterly*, 31(3), 439–465.

Ulwick, A (2005). *What Customers Want: Using Outcome-Driven Innovation to Create Breakthrough Products and Services*. New York: McGraw-Hill.

Urban, GL and JR Hauser (1993). *Design and Marketing of New Products*. Englewood Cliffs: Prentice-Hall.

Urban, GL and E Von Hippel (1988). Lead user analyses for the development of new industrial products. *Management Science*, 34(5), 569–582.

Urban, GL et al. (1996). Premarket forecasting really-new products. *Journal of Marketing*, 60(1), 47–60.

Utterback, J (1995). *Mastering the Dynamics of Innovation*. Cambridge: Harvard Business School Press.

Van Kleef, E et al. (2005). Consumer research in the early stages of new product development: a critical review of methods and techniques. *Food Quality and Preference*, 16, 181–201.

Veryzer, RW (1998). Key factors affecting customer evaluation of discontinuous new products. *Journal of Product Innovation Management*, 15, 136–150.

Veryzer, RWJ (1998). Discontinuous innovation and the new product development process. *Journal of Product Innovation Management*, 15, 304–321.

Von Hippel, E (2001). User toolkits for innovation. *Journal of Product Innovation Management*, 18, 247–257.

Von Hippel, E and R Katz (2002). Shifting innovation to users via toolkits. *Management Science*, 48(7), 821–833.

Vuola, O and A-P Hameri (2006). Mutually benefiting innovation process between industry and big-science. *Technovation*, 26, 3–12.

Wind, J and V Mahajan (1997). Issues and opportunities in new product development: an introduction to a special issue. *Journal of Marketing Research*, 34(1), 1–12.

Yin, RK (2003). *Case Study Research: Design and Methods*. Thousand Oaks, CA: Sage.

USER-PRODUCER INTERACTIONS IN EMERGING PHARMACEUTICAL AND FOOD INNOVATIONS

E. H. M. MOORS*, W. P. C. BOON, R. NAHUIS
and R. L. J. VANDEBERG

Utrecht University, The Netherlands

Introduction

Current pharmaceutical and food innovation trajectories are being confronted with several challenges that revolutionise the way we think about health, and predict, prevent and treat illness. These challenges include demographic changes (ageing population), actual health threats (e.g., climate change, metabolic syndrome, infectious diseases) and rapid developments of radically new technologies, such as biotechnology, nanotechnology and genomics (COM, 2007). Additionally, there is an increased demand of better-informed users for higher added value and more personalised, safe and affordable medicines and foods, which largely improve quality-of-life. All these developments lead to more regulatory hurdles, higher liability pressures, reimbursement restrictions and higher development costs (Atun *et al.*, 2007). Furthermore, the pharmaceutical drug development pipelines are nowadays less productive in terms of real innovative therapeutic compounds (FDA, 2005).

Following these problems and trends, successful translation of basic scientific discoveries into novel medicines and new food products results in an increasingly complex and risky business. Innovation studies show that intensified UPI may tackle some of these problems mentioned and this can increase chances for successful innovations. User needs can be identified and their role in the innovation processes strengthened. Producers are interested in societal acceptance of their products, in access to users' knowledge and in mobilising the creative potential of users. Smits and Den Hertog (2007) distinguish five dimensions on which UPI improves the quality of innovation processes: (1) more effective articulation of social needs;

---

*Corresponding author.

(2) enhanced competitive strength of enterprises; (3) improved acceptance and social embedding of knowledge and technologies; (4) improved learning capacity of society as a whole and (5) enhanced democracy.

While user involvement in innovation processes might be beneficial, it remains unclear how to organise UPI in an effective and efficient way. This counts even more when emerging technologies are involved. Uncertainty and flexibility — inherent to emerging technologies — open possibilities for far-reaching user involvement but at the same time ask for thorough organisation of these UPIs in the face of ever-changing technology specifications, demands and network structures. Furthermore, users do not constitute a monolithic block. Research in UPI heavily depends on the perception of the different roles of users in innovation processes. From the literature, a whole range of perspectives can be derived. In analysing UPI, it is important to distinguish between different types of user-producer constellations.

To study UPI in innovation processes involving emerging technologies, therefore, first of all, a classification of user involvement needs to be made. Such a classification includes a contextualised view of UPIs, i.e., a view that is susceptible to differences in these interactions in different settings and situations. Developing and using such a model might also contribute to suggestions for improvement of the quality of these innovation systems in terms of articulation of social needs, competitive strength, social embedding, societal learning and democratic quality. Therefore, the central research question of this paper is: *How to organise and manage UPI in emerging pharmaceutical and food innovation processes?*

The focus will be on emerging pharmaceutical and food innovations. In both health care and the food sector, companies increasingly anticipate consumer needs and consumer involvement. In the pharmaceutical industry, user involvement has recently been regarded as a way to make the industry more sustainable, i.e., by assisting in questions about rationing of health care expenditure, level of safety required and the validity of medical needs (Atun *et al.*, 2007; Moors and Schellekens, 2008). The latter, amongst others, refers to users, who are asking for life style medicines, and the pharmaceutical industry that sells more comfort drugs with a controversial medical need or for ordinary, mild, personal, or social ailments (Triggle, 2007) instead of drugs for chronic or life-threatening diseases. Concerning the food sector, new products are not only intended to originally satisfy hunger but also to prevent nutrition-related diseases and to increase the physical and mental well-being of consumers (Menrad, 2003).

The next section provides a theoretical overview of user involvement and UPI, and presents a contextualised classification for studying cases of UPI with regard to emerging innovations. Then, the ways to illustrate and deepen important elements of this model are investigated and the applied research methodology is described. Case studies about demand articulation and interactive learning in the pharmaceutical and food sector respectively, provide a dataset for analysis. Next, the results of the case studies are presented and discussed. The paper ends with conclusions and recommendations.

Theory and Classification of UPIs

In general, pharmaceutical and food innovation processes represent science-based innovation trajectories, carried out by a network of interrelated actors, such as universities, research institutes, producers, government and consumers. They are guided by expectations about potential customers and new product innovation, adoption and diffusion. As the majority of new products fail, new product development, especially in emerging technological fields, is a risky endeavour, but essential for the health and survival of a company (Cooper, 1993). The success of new products is improved when there is true added value to the consumer (Griffin, 1996). It is difficult, however, to fully understand user needs and preferences, and to balance them with the strategy of producers to make a product that satisfies users better than competing alternatives. After all, users are not always able to articulate their needs, preferences or wishes, due to the fact that they are not fully aware of all (latent or future) possibilities of a new technology or do not want to share their creative ideas and opinions (Hamel and Prahalad, 1994; Griffin, 1996). Furthermore, identifying opportunities for new products, especially for radical new, emerging products, is quite difficult, as these products "can offer new, unique, or superior solutions to users' needs and can create entirely new markets" (Schmidt and Calantone, 2002). Accordingly, incorporating the "voice of the consumer" in early stages of new product development has been recognised in various studies as a critical success factor (e.g., Griffin and Hauser, 1993; Van Kleef et al., 2005) but at the same time, this voice is difficult to assess.

The traditional economics literature has mostly ignored the dynamic relation between research and product development choices of firms on the one hand and user needs and preferences on the other. In innovation studies, the dynamics of this relation is increasingly perceived as a co-evolutionary process, an institutional interplay in which many heterogeneous stakeholders interact in complex ways. The emergence of new functionalities of a product innovation is a particular aspect of the widening process of co-evolution of a new technology and its users. More or less heterogeneous user groups provide feedback about how new technologies, with varying degrees of flexibility regarding product specific characteristics and uncertainty about potential applications and related ethical, legal and social aspects, match their needs, preferences and performance criteria. These aspects become articulated in demands and interactions between users and producers. Accordingly, users are increasingly recognised as important co-developers of innovations, often developing new functions for technologies, solving unforeseen problems and demanding innovative solutions. This even holds true for science-based innovation processes such as in the pharmaceutical and food sectors (e.g., Lütje, 2003; Verbeke, 2005; Lettl et al., 2006; Urala and Lähteenmäki, 2007).

This paper attempts to bridge the gap between the rather classic linear innovation model for understanding pharmaceutical and food innovation processes, and more recent theorising on innovation systems, characterised by feedback and co-evolution, in which UPI plays an important role. Demands, needs, wishes and concerns of users not only become visible in the end stage but often are articulated throughout the innovation process, for example, in research agendas of firms, wishes of retailers and experiential knowledge of users. Via such interaction and articulation processes, important societal aspects are introduced in innovation processes (Nelson and Winter, 1982; Rip and Kemp, 1998). However, effective involvement of users and other stakeholders by no means is easy and can be improved considerably. The organisation of an effective UPI puts new requirements on the innovation systems in which these firms have to operate. Changes in science and technology as well as the context in which pharmaceutical and food firms have to manage their innovation processes will lead firms to replace their linear innovation model by a systemic, multi-actor model. Taking a holistic perspective of innovation (Atun and Sheridan, 2007), and involving users in innovation processes is an important consequence hereof (Smits and Boon, 2008).

A broad set of disciplines has focussed on the role of users in technology development, ranging from cultural and feminist literature to science, technology and innovation (STI) studies. The latter has recently shown that intensified, and well-designed UPI may increase chances for successful innovations (e.g., Von Hippel, 1988; Lundvall, 1992; Coombs, 2001; Geels, 2002; Smits, 2002; Moors et al., 2003; Oudshoorn and Pinch, 2003; Rohracher, 2005; Smits and Den Hertog, 2007; Boon et al., 2008; Nahuis et al., 2008; Smits and Boon, 2008). Apart from consensus about the (potential) positive impact of users on innovation processes, however, there is not much agreement on questions like how to perceive and manage UPIs in different contexts and/or with different goals.

To develop a more specific and contextualised view on UPI, we have performed a literature review about how different forms of interaction might contribute to the processes of co-evolution in different circumstances (Nahuis et al., 2008). The outcome of the review is an overview of types of interaction. These "types" are inspired by theoretical concepts, such as demand articulation, learning by using, learning by interaction, innofusion, frame sharing and domestication (Teubal, 1979; Rosenberg, 1982; Fleck, 1988; Silverstone and Hirsch, 1992; Bijker, 1995; Boon et al., 2008). They are classified according to a scheme with three axes:

- the phase of technology development

- the level of technological flexibility
- the heterogeneity of the user population

It is widely acknowledged that the characteristics of technology development change along different *phases of technology development* (Collingridge, 1980; Utterback, 1994) with important consequences for the types of UPI that should be employed in these different phases (Rip and Schot, 2002; Stewart and Williams, 2005). Somewhat simplified, we distinguish between an early phase when actors are building up a protected space, and a later phase in which the technology enters the wider world. Yet, the phase of technology development is the least relevant axe in the current paper, because we focus on emerging technologies, which, by definition, are in an early phase of development.

Regarding the *flexibility of the technology*, we use Fleck's (1994) notion of configurational technology. Configurational technologies are built up from a range of components to meet the specific requirements of particular user organisations. A technology is called flexible when different configurations with different performance characteristics are reasonably possible. If not, we speak of specific technology.

Turning to the *heterogeneity of the user population*, we note that there are several sources of heterogeneity: different users have different capabilities and knowledge bases (Akrich, 1995), user contexts are often unique as a consequence of contingent historical developments (Fleck, 1994; Garrety and Badham, 2004), and there may be different kinds of users of the same technology that have different needs and concerns (e.g., medical professionals, nurses, patients and hospital administrators in case of medical technologies) (Oudshoorn and Pinch, 2003). Although "users" is understood as an emerging category that is amenable to change (Martin, 2001), there are nonetheless important differences across and within technology fields as to its degree of heterogeneity. For example, users of apartments appear to be much more homogeneous than users of addiction treatments (Franke and Von Hippel, 2003). We speak of homogeneous user populations when many users are satisfied with the same standard product. Users are heterogeneous when user contexts are very unique or complicated, or when users have conflicting interests in the development of certain technologies.

These three dimensions constitute a classification scheme for UPI. Theoretical concepts for UPI are derived from a number of relevant bodies of literature, namely evolutionary economics, constructive technology assessment, social construction of technology, semiotic studies and cultural studies.[1] The literature review shows

---

[1] Clearly, this review could be extended with concepts from other bodies of literature, such as democratisation theory, marketing research and risk research. For an overview of the literature about science, technology and democracy, see Nahuis and Van Lente (2008) and Bucchi and Neresini (2008).

how the requirements to UPI vary with the phase of development, the flexibility of technology and the heterogeneity of users. Table 1 presents the results of this review.

This classification scheme becomes especially useful when it is not only clear which types of UPI are relevant in what circumstances, but also what the main conditions and operations of these types are. We consider this scheme as an agenda for future research into the whole spectrum of UPI and we invite other scholars to join this endeavour. As a start, this paper somewhat arbitrarily zooms into the two emphasised types of UPI and explores the purposes, conditions, mechanisms and outcomes of *demand articulation* in circumstances of specific technology and homogeneous users, and *interactive learning* in circumstances of specific technology and heterogeneous users.

In both demand articulation and interactive learning, a distinction can be made between first-order and second-order learning (e.g., Sabatier, 1987; Cohen and Levinthal, 1990; Boon et al., 2008). If the technology is adapted to fit the user environment, first-order learning takes place. It refers to instrumental knowledge concerning specific knowledge for solving a problem within an existing framework. The aim is to make existing user preferences explicit, which can then be translated into design requirements. Second-order learning is conceptual knowledge related to the framework itself and the shared vision amongst users in this framework. In second-order learning users experiment with new ways of using artefacts, allowing them to question established assumptions about functionalities of technologies (underlying assumptions), cultural meanings, user practices and preferences.

Research Methodology

According to Table 1, *demand articulation* is important in several circumstances. Demand articulation is defined by Boon et al. (2008) as "an iterative, inherently creative process in which users try to unravel preferences for and to address what they perceive as important characteristics of an innovation". Demand articulation takes place when preferences of users are made explicit, in such a way that it prompts other actors to (re)act. In many cases, users do not yet have precise demand requirements and a clear view of relevant product attributes. Users' needs and possible alignments with technological opportunities cannot be discovered *ex-ante*, as scholars

---

For a discussion about incorporating consumer preference in product development in the marketing literature, see Schmidt and Calantone (2002), Griffin and Hauser (1993) and Van Kleef et al. (2005). For a comprehensive discussion about the difficult relation between value pluralism and risk assessment methodology and its implications for public participation, see Renn (1998), Stirling (1998) and Fischhoff (1995).

Table 1. A classification of UPI types.[a] The emphasis on two types of UPI is added in order to position our case studies.

		User population		
		Homogeneous		Heterogeneous
		Protected space	Wider world	Protected space	Wider world
Technology	Specific	**Articulation of needs, demands, concerns and opportunities**[b]; mobilisation of creativity of potential users	Learning how to integrate technology in practice; exploring whether real use is conform expectations; reducing ambiguities of technology	Broadening stakeholder participation; developing procedures for legitimate user representation; **exchanging knowledge and debating values and visions**[c]; articulation of demands, concerns and opportunities; developing structures and codes for frequent interaction	Learning how to integrate technology in practice; teaching users; reducing ambiguities of technology; learning how technology performs in real life circumstances over time
	Flexible	Frequent articulation of needs, demands, concerns and opportunities; mobilisation of creativity of potential users; developing structures and common codes for frequent interactions; in some cases predominantly users initiate or press innovation	Learning how to integrate technology in practice; additional/ complementary demand articulation; exploring whether real use is conform expectations; reducing ambiguities of technology	Broadening stakeholder participation; assessing existing visions and technological frames; sometimes users construct prototypes	Learning how generic technology may help solving specific practical problems; adapting technologies to local circumstances; learning how technology performs in different circumstances over time; learning how users could further innovate products; additional/ complementary demand articulation; users buy components to construct prototypes of new products

[a]Adapted from Nahuis *et al.* (2008).
[b]Further referred to as demand articulation.
[c]Further referred to as interactive learning.

like Teubal (1979) and Clark (1985) stressed. They have to be constructed and negotiated in a process of mutual articulation and alignment of demand and supply. In this process, the role of intermediary organisations, such as mediators, advocacy groups or spokespersons, is relevant.

Although demand articulation is important in several circumstances, we selected a case with relatively homogeneous users and a rather specific technology in accordance with our interest in emerging pharmaceutical technologies. In a pharmaceutical discourse, demand is fundamentally linked to disease. Defined by a common disease, patients constitute a relatively homogeneous population. Disease does, however, not fully determine demand; there are several related aspects of demand that remain to be articulated in relation to the characteristics of an innovation, such as the safety and efficacy of drugs. Pharmaceuticals are generally specific, because they are R&D intensive and composed of components that cannot be configured differently without losing essential performance characteristics.

The second type of UPI under study here is *interactive learning*. Users and producers are experimenting with the new technology and interactively learn on a variety of dimensions. Then alignment takes place (Rosenberg, 1982; Lundvall, 1988). Interactive learning is important when information and knowledge are tacit and difficult to communicate. This is generally the case in early phases of technology development (Vandeberg and Moors, 2007). Moreover, learning by interaction is crucial in any case where frequent interaction is required, i.e., when actors have to rely on one another's expertise. Such reliance exists among others in the combination of specific technology and heterogeneous users, because such technologies are bound to fail without compromise or alignment of the needs and concerns of heterogeneous users (Nahuis et al., 2008).

We selected a case from the field of nutrigenomics as an example. Users are heterogeneous in this case, since they come from science, industry, government and the public, bring in their own expertise and express their own needs and concerns. The technology can be characterised as specific, given the knowledge intensive outcomes of interactive learning. The sort of technology-as-knowledge, which interactive learning currently helps developing in nutrigenomics (about gene-nutrient interactions) is not flexible in the sense that it can be rather freely recombined to get different results; recent insights at best reveal specific gene-nutrient relations.

We should also note that both the pharmaceutical and the food innovation system face new challenges in the era of genomics with implications for the classification presented here. The emergence of genomics is said to initiate three trends: (i) from cure to prevention (more emphasis on diagnosis), (ii) from relatively homogeneous user populations (patient groups) to personal needs (a unique set of predispositions) and (iii) from specific products to customisable therapies/diets. Nutrigenomics researchers, for example, state that the purposes at the horizon of the research landscape are to capture human genetic diversity with nutrigenomics and serve individual needs with tailored diets and therapies (Kaput *et al.*, 2005). Similar expectations are expressed for pharmacogenomics, although patient stratification in substantial groups will probably still be preferred to pure individualisation of demand (Royal Society, 2005; Trusheim *et al.*, 2007). Nevertheless, future cases should perhaps be positioned differently in the classification scheme (Table 1) as these trends are arguably directed towards the lower-right corner of the scheme. While such visions still mainly figure at the level of promises and expectations, the case studies presented here could function as "baseline studies" to be confronted with such trends in order to explore the implied shifts in innovation systems. Such an exploration yields important insights for innovation management and policy in general and the future organisation of UPI in particular.

The case studies in this paper are presented to *illustrate* and *deepen* the classification of UPI types; they analyse the conditions and mechanisms for demand articulation in current pharmaceutical and interactive learning in current food innovation systems (Boon, 2007a,b; Vandeberg and Moors, 2007; Boon *et al.*, 2008; Nahuis *et al.*, 2008; Vandeberg, 2008a,b). These studies focus on intermediary organisations and consortia, respectively. The reason for this focus is that these are the settings where such articulation and interactive learning processes typically occur (Kaput *et al.*, 2005; Boon, 2007a,b; Vandeberg, 2008a,b). Therefore, this paper not only explores the purposes, conditions, mechanisms and results of interactions in these different settings, but also allows for a comparison of the relative merits of these different ways of organising UPI.

However, there are some limitations to this comparison, which follow from methodological considerations in the case studies. For investigating demand articulation, first, an *agency* perspective is adopted. The main reason is that intermediary organisations are important agents in the process of demand articulation and following their activities is a proper heuristic for identifying needs and concerns and for examining the role of articulation in innovation processes. Interactive learning in consortia, in contrast, is studied from a *structure* perspective, because consortia are brought into being due to the fact that they are believed to form an adequate structure for learning. A related conceptual difference is the inclusion of time as a variable. Whereas demand articulation from an agency perspective necessitates following an actor over time, understanding interactive learning from a structure perspective predominantly requires mapping spatial, cognitive and cultural aspects of interaction.[2] In other words, demand articulation is studied following a process-based model, whereas interactive learning makes use of a variance model (Poole *et al.*, 2000).

---

[2] See Boon *et al.* (2007a,b, 2008, In Press), Vandeberg & Moors (2007), and Vandeberg (2008a,b) for further clarification and argumentation of research design decisions.

Studying demand articulation implies the importance of following stakeholders in long-term debates, the activities or "events" they participate in, and the related demand statements they make in the context of these events over time. To analyse the time-ordering and the changes in content-related issues such as demands in a structured way, the "event history analysis" method was used as devised by Poole *et al.* (2000). Examples of such events include conferences, published reports, meetings and deliberations. We explicitly followed one patient advocacy group, studying how they formed and expressed their demands and what the content of these demands was.

The data were obtained from the archives of the patient advocacy group, including minutes of meetings (board meetings but also those of committees), letters, reports and evaluation. Other data sources include more open-access ones. In addition, we conducted interviews with several representatives of the patient group, and substantiated these with interviews with representatives from organisations that frequently interacted with the patient organisation. These interviews were primarily meant to clarify archival information, provide the easiest inroads to the data, and by asking "why-questions" uncovering the underlying assumptions behind the demands that were found.

For the study of interactive learning in emerging technologies, a framework was developed that incorporates structural aspects, like proximity/distance between stakeholders on several dimensions, as well as process variables like knowledge flows, network formation and the role of a prime mover and intermediaries therein. Indicators for these variables are identified in order to distil and analyse relevant data.

General information on (the emergence of) consortia, the stakeholders within consortia and policy surrounding these consortia was derived from consortia and stakeholders' websites and complemented with internal and policy documents (publicly available). Scientific articles and patents were used to assess the outcome of the consortia and news articles and other (e.g., websites) publications reporting about the consortia were used as complementary material. The interactive learning process was assessed through interviews with consortia stakeholders. The most knowledgeable respondents were selected, based on their participation in projects and overall understanding of the consortium. Some concepts could also be assessed quantitatively. UCINET6[3] was used for co- and cross-citation analyses, i.e., cognitive proximity, and Google Maps[4] to analyse the travelling time, i.e., geographical proximity, between the stakeholders in a consortium.

---

[3] http://www.analytictech.com/downloaduc6.htm.
[4] http://maps.google.com/

Results

This section presents the results on demand articulation mechanisms in emerging pharmaceutical innovations and interactive learning conditions in emerging food innovations, respectively in order to illustrate the developed classification of UPI types, i.e., phase of technology development, flexibility of technology and heterogeneity of users. The first part focuses on demand articulation processes of the Dutch Neuromuscular Diseases Organisation VSN (Vereniging Spierziekten Nederland), a patient group that works, amongst others, on emerging drugs for Pompe disease. The second part illustrates interactive learning conditions in the case of emerging nutrigenomics developments in the Dutch Nutrigenomics Consortium.

Demand articulation of a patient organisation in the context of Pompe drug development

The VSN is a patient organisation that was founded in 1967 by a group of parents with children who had a neuromuscular disease. The organisation shared information with other patient organisations for the need to support information provision to patient members, and set-up mutual help services. The VSN dealt with several promising, emerging technologies, such as gene and stem cell therapy, in their role of information translator. Nevertheless, what makes the VSN quite unique in the Dutch health care context, is their contribution to research and development of diagnostics and therapy for neuromuscular diseases.

Developments in pharmaceutical science and technology spurred the patient organisation to pursue other directions of action. Already in the 1970s, the VSN recognised that many neuromuscular diseases are genetically determined. The VSN's focus on research pushed to follow the train of scientific events: the location of the disease susceptibility gene should lead to the identification of this gene. In turn, this discovery contributed to finding a diagnostic tool and producing more information about the natural course of the disease. Subsequently, the products of the gene (proteins) were identified, which might be the starting-point for therapy. In order to control part of these developments, the VSN played a part in raising funds, and supporting scientific work through funding international research networks. The first disease area in which this "train of scientific events" model was used was Duchenne muscular dystrophy. But Pompe disease was the most prominent case in which the VSN became involved and articulated demands. This case led to the development of a therapy in which a deficient enzyme in Pompe patients was replaced, i.e., enzyme replacement therapy.

Pompe disease (glycogen storage disease, acid maltase deficiency) is a rare, genetic, metabolic disorder causing progressive skeletal muscle weakness. After the discovery and cloning of the gene that is responsible for deficiency in the production of the enzyme acid alpha-glucosidase that causes Pompe disease, at an Erasmus MC research group in 1990, this group looked for possibilities to upscale the production of enzymes using this genetic knowledge. It chose to ally with the Dutch biotechnology company Pharming that worked with transgenic animals and had produced a transgenic bull. Genetic modification of animals for producing enzymes had become contested in the political debate at that time. Several actors, including the legislator, called for stricter regulation. The VSN tried to voice the anxieties of patients and the biomedical world, but their demands remained rather marginalised. Later, the VSN condemned promotional actions by an animal rights group and had considerable more success.

The VSN also tried to influence other aspects of the drug R&D process, such as the set-up of phase I/II clinical trials, demands about compassionate use of the Pompe drug during phase III-clinical trials (i.e., the use of the drug before regulatory approval by very ill patients who have no treatment alternatives), the approach and estrangement later on of the two companies involved (Pharming and Genzyme), the reimbursement of orphan drugs, the drug's approval for late-onset Pompe patients, and putting newborn screening for Pompe on the agenda. Part of these advocacy efforts was actively channelled through an international spin-off, the International Pompe Association (IPA), founded in 1999.

When following the events in which the VSN was engaged in over time, the articulation of demands could be analysed. The demands of the VSN became increasingly concrete. For example, regarding the aforementioned animal rights groups the VSN converged its opinions and sharpened its argumentation. Although the VSN was following the events that were topical at various moments of the event cycle, the organisation mostly took a proactive stance towards Pompe drug R&D. The influence on the drug R&D process was substantial, but only in terms of easing the communication between actors and by creating the right conditions, e.g., stimulating clinical trials and reimbursement. Although the VSN — and IPA — had influence on the speed of the innovation process, addressing contested issues and filling-out of the innovation process and the implementation of the drug, they did not have an impact on which technological alternative should be chosen to enter the market as a drug. For this, the interests and the power of the companies involved, Pharming but later also Genzyme, were just too great.

The VSN was thus engaged in first-order learning, i.e., converging demands, problems and ideas. The demands focussed on drugs based on the enzyme replacement therapy principle, the speedy development of this drug, and related boundary conditions, such as reimbursement. The VSN and the patients it represented were neither in the position to be particular nor powerful enough to impose other treatment alternatives. Moreover, the VSN also learnt on a second-order level by developing their underlying assumptions. The assumptions include positioning themselves

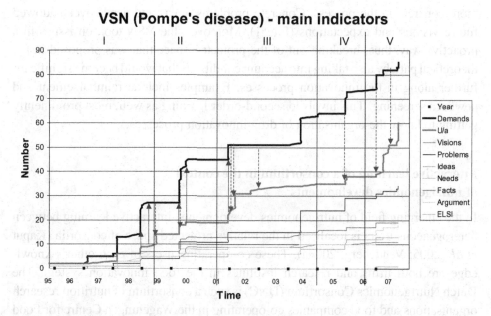

Fig. 1. Cumulative amount of events and statements in the event cycle on ERT for Pompe disease. The main categories are demands (subdivided into demand categories, such as visions, problems and needs) and underlying assumptions (u/a), which are interlinked through arrows.

towards other actors, e.g., by stressing the importance of the "natural alliance" between Erasmus MC and the VSN; constructing the role it played of, amongst others, proactive broker and representation of patient interests, as a model for other neuromuscular disease associations to follow; regarding the timing of advocacy activities; and trust that the VSN puts in other parties. The latter refers to the fact that the VSN does not trust other parties on face value anymore (some interactions caused the VSN to re-define its faith in others), but always requires some base of evidence (Boon, In Press, 2008). Finally, these first- and second-order learning processes also influenced each other (see Fig. 1) by legitimising demand articulations using underlying assumptions (upward arrows), and evaluating underlying assumptions following demand articulation (downward arrows).

To conclude, this case showed that the VSN was much engaged in representing their patient members towards other actors by (1) putting the demands from these members on their agenda, (2) synthesise these demands using its own underlying assumptions, and subsequently (3) express and (4) evaluate their demands to other actors. These other actors, including companies, research groups and government agencies, were influenced by these advocacy efforts and resulting demands. This user involvement led at least to facilitating the Pompe drug innovation process, in which the VSN constructed and articulated their demands, and made these demands more concrete in the process. Concrete problems, ideas and needs overshadowed future visions and expectations (Fig. 1). Moreover, the VSN took on issues in a proactive way, thus, breaking out of the protected space that was proposed in the theoretical part, by also taking into account conditions that would become significant further along in the innovation processes. Examples include reimbursement and newborn screening. This involved second-order learning as well, most prominently getting to know the organisation of drug innovation processes.

Interactive learning of a consortium in the context of nutrigenomics developments

In the emerging field of nutrigenomics developments, interactive learning between heterogeneous users is localised in the relative protected space of consortia (Kaput et al., 2005; Vandeberg, 2008a). These co-operating users of each other's knowledge are both firms and research institutes in the food innovation system. The Dutch Nutrigenomics Consortium (DNC) is such a consortium of nutrition research organisations and food companies co-operating in the Wageningen Centre for Food Sciences (WCFS) and medical/genetics research organisations co-operating in the Centre for Medical Systems Biology (CMSB) that joined their forces to form a nutrigenomics research collaboration.

The DNC was operational for the period 2003–2007. The outcome of the interactive learning process regarding emerging nutrigenomics developments is represented in scientific knowledge (i.e., first-order) and shared visions (i.e., second-order) in the DNC consortium. The shared vision in the DNC is the result of an "attuning process" between the top-down mission at the Dutch governmental level, which co-financed the DNC, and the bottom-up emergence of a shared vision within the DNC (Fig. 2). Although the shared vision slightly emphasised the unfolding of metabolic stress as a topic and conservatively predicted future applications of food components, during the scientific nutrigenomics research itself a shift within the shared vision became visible. Through the scientific research, the knowledge flows within the DNC and the resulting outcome (e.g., scientific articles), it was found that nutrigenomics is far more complex than thought during the formation of the DNC. Therefore, the shared vision and the entailed expectations shifted from *solving* the metabolic stress/ syndrome, with concrete food products available in a wider world, to *understand* the basic scientific principles responsible for or leading to the metabolic syndrome.

The (shifting) shared vision and the scientific knowledge outcome of the DNC are the result of interactive learning, which encompasses characteristics and conditions of the interactive learning process. Network formation, the role of prime movers and intermediaries, and the flow of knowledge are important interactive learning

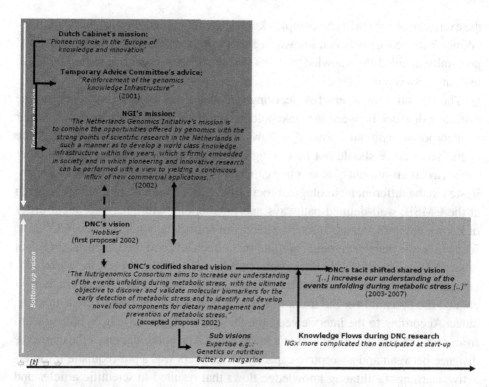

Fig. 2. Construction of a shared vision (second-order learning) and influence of scientific knowledge (first-order learning).

characteristics when emerging technologies, such as nutrigenomics are concerned. These interactive learning characteristics are influenced by various proximity conditions, including geographical, cognitive, regulatory, cultural and organisational proximity (cf. Vandeberg and Moors, 2007). In the case of the DNC consortium, several conditions influencing the interactive learning process are visible.

It is assumed that firms located in areas with other firms have better innovation performance than more isolated or distant firms: organisations benefit from being located close to other organisations (Weterings, 2006). In science-based emerging technologies complex knowledge is exchanged between knowledge users. Often, this knowledge is tacit (Malmberg and Maskell, 1999; Doloreux, 2004) or it is so complex that it needs a tacit explanation (Howells, 2002). The exchange of tacit knowledge between users is enabled through face-to-face interaction (Feldman, 1994; Gertler, 2003). Geographical proximity facilitates face-to-face meetings, and as such geographical proximity is an important condition for tacit knowledge exchange and interactive learning (Feldman, 1994; Gertler, 2003). During the construction of the DNC and especially within the research projects, the close geographical co-location of the stakeholders within the Netherlands enabled the exchange of both tacit and complex knowledge (e.g., regarding scientific nutrigenomics knowledge) between knowledge users in the DNC. As such, geographical proximity enabled the knowledge flows within the DNC which lead to interactive learning between the users.

The potential for interactive learning between users is also influenced by the cognitive distance between the stakeholders and their absorptive capacity. In order to incorporate external knowledge flows into one's own knowledge system, the cognitive distance should not be too big (Boschma and Lambooy, 1999). In other words, a certain amount of absorptive capacity is needed. Cognitive distance is manifested in the different technological foci of DNC stakeholders: In medical research at the CMSB, well-defined pathways and interactions with pharmaceutical drugs of known composition are studied, whereas in nutritional studies of the WCFS the focus is on the pluriform, direct and indirect effects of food on homeostasis (Kaput et al., 2007). Therefore, there is some cognitive distance between the stakeholders with a nutritional or medical background in the consortium. The distance itself created the potential for the users to learn, if they would be able to cross this distance. According to the interviewees, the DNC researchers have been able to learn from each other and use each other's knowledge. The combination of the cognitive distance between and absorptive capacity of the users was a pre-requisite for interactive learning, facilitating knowledge flows that resulted in scientific articles and a (shifted) shared vision. In those instances in which the users were unable to cross the knowledge gap among them, intermediaries functioned as translators or knowledge brokers (Geurts, 1993) between complementary users that are at the outer ends of the nutrition-genomics spectrum. Within the DNC, scientific knowledge was translated through boundary objects (Star, 1989): mouse and human genetics models were used which were both familiar to nutritional scientists (at WCFS) and medical/genetics experts (at CMSB). As such, these boundary objects could be regarded as intermediaries. This enabled the users to discuss and exchange knowledge from a mutual starting point. At the same time, translations from basic science to possible applications were organised in exemplification projects. These made it possible to apply knowledge more often for different purposes and to explore future applications in the wider world of nutrigenomics products.

Regulatory proximity at the innovation system level reduces uncertainty about return on investment (Autio et al., 2008) and stimulates research organisations, which probably would otherwise not endeavour on this research, to co-operate in risk bearing, costly long-term projects and to form networks. The users in the DNC in the Netherlands are all subject to the same regulations at the food innovation system which are directly related to the EU directives, i.e., the EU Novel Food regulation and proposal for a regulation of the European Parliament and of the council on nutrition and health claims made on foods (COM2003/424). These regulations encompass the foresight that nutrigenomics products will be allowed onto a protected market space in the EU. As such, these regulations granted some future return on investment and stimulated the DNC stakeholders to form a network and to perform co-operatively costly, risky science-based nutrigenomics research.

Uncertainty about co-authorship and IPR might hinder the free flow of knowledge within a research consortium that is dependent on this knowledge interchange. Regulatory proximity at the network level, in the form of contracts and mutual non-disclosure agreements (NDA) of knowledge outside the consortium and IPR, reduce the risk of unwanted spill-overs. As such, these agreements encourage knowledge flows. Therefore, the organisations within the consortium found it important to make clear arrangements regarding IPR (even though IPR arrangements might not be economically beneficial (Dosi et al., 2006)). Before the WCFS and CMSB joined their forces for nutrigenomics research, they had their individual patenting mechanisms. Discussion during the formation of the DNC resulted in a mutual agreement that satisfied the CMSB and the WCFS. IPR arrangements prevented unwanted spill-overs because it was clear who owns which knowledge and under which conditions others could use this knowledge (e.g., through patents). Unwanted spill-overs were prevented further by the agreement not to conduct similar research simultaneously outside the DNC. So, the users knew that their knowledge would stay within the consortium and it was clear under which conditions who would be involved in publications and possible patents. Therefore, the users, tended to let their knowledge flow more freely. At the same time, the users have to trust each other that they will not misuse complementary knowledge for their sole benefit. This is especially important because of the ubiquitous character of tacit knowledge. Tacit knowledge flows more easily in the presence of shared trust (Malmberg and Maskell, 1999). The main difference between the different organisations (i.e., universities, research institutes and companies) within the DNC consortium is their different focus on outcome, resulting from the different underlying incentive structures. The differences between the foci of universities and companies on publications and patents respectively, might lead to a conflict: if a finding has been published (in a scientific journal) it can no longer be patented because it is impossible to trace the originality of the idea by the patent applicant.

Cultural proximity is related to these different underlying incentive structures for science and industry, which might induce reluctance to knowledge sharing and could block knowledge flows and interactive learning (Dasgupta and David, 1994; Frenken and Van Oort, 2004). Although the stakeholders within the DNC already agreed on NDA and IPR in the mutual agreement, a culture of shared trust was very important. For the collaboration in the DNC, this trust was partly based on earlier experiences of the stakeholders with each other. At the same time, due to the emerging, pre-competitive phase of nutrigenomics research, no rivalry existed between the knowledge users in the DNC that could lead to an offensive situation, which would also be prevented by the IPR and NDA agreements in the mutual agreement.

Finally, the "production" and co-ordination of complex knowledge is determined by the organisational proximity. Organisational proximity encompasses the flexibility that enables individual pursuit of knowledge goals, and co-ordination that enables the combination of complementary knowledge flows (Boschma, 2005). The DNC was set up as a virtual network in which nutrigenomics research was organised around topics and *"closely interacting work packages [...] headed by WP leaders, who are responsible for scientific leadership and project management of their WP"* (DNC). Within the work packages (WPs), several organisations could participate. Each WP was managed individually by a WP leader, which resulted in flexibility at the WP level. From the beginning, these activities were co-ordinated through the DNC as arranged during network formation. This also entailed standardisation that would make data exchange and knowledge flow between WPs easier.

To conclude, interactive learning processes did take place in the DNC consortium with increased scientific knowledge and a shared vision as concrete outcome. The shared vision shifted: increased understanding of the underlying mechanisms of the metabolic syndrome called for studying these mechanisms further before turning to developing concrete products for the wider world. By studying the influence of proximities on network formation, prime movers, intermediaries and knowledge flows, the interactive learning processes of specific emerging nutrigenomics developments regarding metabolic stress in the protected space of the DNC consortium have been unravelled. The results show that EU regulations reduced the future return on investment, which encouraged the knowledge users in the consortium to form a network and embark on nutrigenomics research. Within this network, the mutual IPR and NDA arrangements prevented unwanted spill-overs, which facilitated the free flow of knowledge. Knowledge flows were further stimulated by a culture of shared trust. The cognitive distance between the heterogeneous knowledge users in the consortium was dealt with by boundary objects, which enabled them to bridge knowledge lacunas based on mutual starting points. The geographical co-location enabled face-to-face interactions between the users, in which tacit and complex knowledge was exchanged. The set up of WPs granted both flexibility and co-ordination of the complementary knowledge in the DNC.

Concluding Remarks and Discussion

The aim of this paper is to understand *how to organise and manage UPI in emerging pharmaceutical and food innovation processes*. These insights could lead to recommendations for improving the quality of these processes in terms of articulation of social needs, competitive strength, social embedding, societal learning and democratic quality. The paper focuses on developing a classification scheme for UPI, in which various theoretical concepts for types of UPI are depicted. It is argued that the requirements for UPI vary with the phase of technology development (in protected space or in wider world), the flexibility of technology (specific vs. flexible) and the heterogeneity of user populations (homogeneous vs. heterogeneous). This classification scheme offers a potential base for evaluating and improving the organisation and management of UPI in innovation processes, because it draws attention to the relation between relevant types of UPI in a certain case and the wider context of the case. After delineating their case, innovation managers are encouraged to consider a particular set of interaction possibilities. To fully realise the management potentials, however, two points need to be taken into account: (1) investigating how particular cases indeed fit into the categories of the scheme and (2) investigating how, once a particular type of interaction is considered worth pursuing, the interaction should actually be organised. Both these points are addressed in this paper, albeit partially, and are discussed below.

First, in order to investigate whether the classification scheme is robust and in which circumstances, this paper zoomed into two types of UPI, i.e., processes of demand articulation and interactive learning. The case of demand articulation by the Dutch patient organisation VSN indeed matched circumstances of a specific technology (i.e., enzyme replacement therapy for Pompe disease) and a homogeneous user population (Pompe disease patients). However, the case was not restricted to the protected space. Quite exceptionally, the VSN already was proactively involved in the drug R&D process and clinical trials. But it was also, and less exceptionally, involved in articulation activities related to topics that need to be addressed once technologies enter the wider world, like drug approval, reimbursement and newborn screening. According to the classification scheme, interactions should then increasingly aim at learning how to integrate technology in practice, exploring whether real use conforms expectation, and reducing ambiguities of the technology. The case study shows that some of these functions, notably realising expectations and reducing ambiguities, are already anticipated in demand articulation processes. Demand articulation in an early phase can thus contribute to improved societal embedding of new drugs.

In the case of interactive learning in the DNC, we described the circumstances in terms of a specific technology (i.e., nutrigenomics research for metabolic stress) and heterogeneous users. Especially the latter characterisation requires further discussion. Users were defined as users of knowledge. This turned any participant of the consortium into a user as they participated to learn from each other. We argued that this is a legitimate conceptualisation of the user when talking about interactive learning, especially in the case of emerging technology where predominantly scientific and technological actors interact about scientific knowledge, and both concrete applications and actual (end) users are still absent. The classification scheme thus proves to be relatively independent from how producer, object and user are reciprocally defined. However, this conceptualisation of users is not without repercussions for the position of the case in the classification scheme. If the knowledge developed in the DNC in due course leads to the development of more or less concrete genomics-based technologies or products, the user of this technology might not be the same as the user of the underlying knowledge that we studied. As a consequence, the case will shift within the classification scheme for methodological reasons.

What is more, apart from methodological reasons, cases can also move through the scheme as a result of trends and developments within innovation systems or society at large. The possible transition of the health care system towards enhanced genomics-based diagnosis, delineation of individuals as a unique set of predisposition, convergence of food and pharmaceuticals, and personalised therapies and diets might trigger or necessitate a fundamental reform of both the pharmaceutical and the food innovation system. This raises interesting new questions. If these sectors indeed move towards heterogeneous user populations and flexible technology, then the classification scheme suggests that UPI types, such as innofusion and user innovation become increasingly important. But what would these concepts, originating from IT-related innovation studies, actually mean in a health-related context? What are their mechanisms and what conditions do they pre-suppose? What would, for example, be the role of intermediaries in these types of UPI and does proximity still matter as much? Although answers to these questions are beyond the scope of this paper, we think we have at least coined the concepts with which such questions can be addressed.

The *second* need for realising the potential of the classification scheme to become a useful toolbox for innovation management and policy is to get a grip on the conditions and forms of organisation of UPI. The analysis of the UPI processes, mechanisms, conditions and the outcomes of demand articulation and interactive learning give important points of attention for improving emerging innovation processes, both in the protected space and in the phase of the wider world.

Building on the two exemplary cases, the most important conclusions regarding organising and managing various types of UPI in emerging pharmaceutical and food innovation are:

Firstly, regarding the UPI type, *demand articulation,* the conditions of demand articulation in a patient organisation include problem-, vision- and agenda-setting, demand synthesis by underlying assumptions, expression and evaluation of demands with other actors, all leading to facilitating emerging pharmaceutical innovation processes. In the context of the role of demand articulation processes in pharmaceutical R&D processes, the Dutch Neuromuscular Disease Association (VSN) articulated its demand on several issues at several stages of the Pompe drug development process, including compassionate use, expectations of different stakeholders *vis-à-vis* certain innovations, etc. The results showed that representative intermediary user/patient organisations, such as the VSN, can have a beneficial and steering impact on pharmaceutical innovation processes.

Secondly, with regard to the UPI type, *interactive learning,* the characteristics for interactive learning in a consortium include stimulation by geographical, cognitive, cultural, organisational and regulatory conditions, finally leading to (changing) shared visions and a collective scientific output, leading to "improved" nutrigenomics innovations.

For consortia to interact and learn from each other, we found in the case study of the DNC that a certain amount of cognitive distance and absorptive capacity between the stakeholders turned out to be crucial for innovation in science-based food innovations (e.g., nutrigenomics), as are trust and mutual agreements in case of conflicting interests.

In the pharmaceutical innovation system, intermediary organisations, such as a patient organisation, enter through various first- and second-order converging articulation mechanisms (compassionate use, organising scientific workshops, clinical trials and newborn screening) into ongoing developments — when they try to influence drug research and development pipelines. The intermediary organisation also plays a role in toning down exaggerated expectations of the new technology. However, the presented VSN case should ideally be posed against cases of other intermediary (patient) organisations, because this category of actors is heterogeneously setup. In the food innovation system, an intermediary role is often fulfilled by bridging objects like mouse models.

Zooming in on two types of UPI, namely demand articulation and interactive learning, recommendations could be presented for policy makers and various user groups on how to organise user involvement in emerging innovation processes in the pharmaceutical and food sector in a more effective and efficient way, by focussing on the described mechanisms of demand articulation and interactive learning.

Regarding demand articulation mechanisms, articulation of needs, demands, problems and opportunities/expectations are important by clarifying agenda-setting practices. Bringing in the vision of other actors leads to mobilisation of the creative potential of prospective users. Now the developed Pompe drug is on the verge of stepping into the wider world, the intermediary user organisation plays an important role in learning how to bring the new enzyme replacement therapy technology to all its users, exploring whether the real use conforms expectations and guidelines. What is more, looking for opportunities to broaden the indication area and the spectrum of users (patients with early-onset, juvenile and late-onset Pompe disease) might even turn the case into one of heterogeneous demand, meaning that additional attention should be paid to integrating technology in various practices and to learning by using in these practices.

Regarding mechanisms of interactive learning, it became clear that the nutrigenomics developments are still in the protected space phase, with shifting shared visions about potential nutrigenomics applications and steadily increasing scientific output. Within the studied consortium, broadening of user participation took place via network formation and mechanisms of absorptive capacity (cognitive proximity), and exchange of knowledge (presence of knowledge flows). The geographical, organisational and regulatory proximity conditions could facilitate structures for emerging technology development, and codes and networks for frequent interaction between complementary stakeholders. Demands, concerns and opportunities are amongst others articulated by (shifting) shared visions.

Table 1 could be further developed as a toolbox with respect to the management of UPI in emerging pharmaceutical and food innovation systems, helping various stakeholders in organising involvement of users in emerging innovation processes in a more effective and efficient way.

Summarising, this paper shows that systemic differences best account for how UPI should be organised and managed in emerging pharmaceutical and food innovations. One implication for innovation management is that successful forms of UPI, such as demand articulation and interactive learning, cannot be copied from one case to another without taking into account the institutional environment of the innovation system (the development of an emerging technology being co-shaped by a number of geographical, organisational, regulatory and cognitive systemic conditions) and the nature and role of intermediary organisations. Producers of emerging pharmaceutical and food technologies should stay focused on other stakeholders, consumers and patients with specific needs, via intermediary user organisations, consortia or other forms of user-producer linkages. This will help in developing products in a co-evolutionary way towards better societal embedment.

This paper showed that organised UPIs, for example via intermediary user organisations such as the VSN, and via consortia, such as the DNC, are important tools and opportunities for demand articulation and interactive learning involving patient organisations, researchers and private and public organisations.

The results presented in this study should be regarded as tentative due to the exploratory nature of the research carried out. Regarding the classification scheme in Table 1, only two situations were studied, i.e., specific technology in a protected space/wider world with homogeneous users, and specific technology in a protected space with heterogeneous users. As discussed, we endeavour a large program in which we are able to investigate the conditions and processes of UPI in many different circumstances, starting with the food and pharmaceutical innovation system. This paper presented the first results in the context of the integrative classification frame. Yet, further research should be conducted along various lines of research. Firstly, similar case studies should focus on other types of UPI and/or in different kinds of cases in order to more fully realise the practical relevance of the classification scheme. Secondly, the same research can be conducted in other high-income countries. Then, the results could be combined with those from the Netherlands to carry out a more reliable international comparative study and to emphasise the role of institutional and cultural differences in the organisation of UPI. Thirdly, when in future the first nutrigenomics innovations are actually introduced to the market (stepping in the wider world), nutrigenomics development trajectories can be compared with pharmaceutical trajectories on a case-study basis, and the mechanisms of UPIs could be studied more in depth. Fourthly, further research should be done to assess the viability of a trend towards personalised therapies and diets. The results of current studies could be used as a baseline for such assessment. Finally, future research should also provide insight into the extent to which the results of this study are useful in other sectors with emerging technologies, such as IT, nanotechnology and pharmacogenomics, thereby increasing the understanding of the role of UPIs in innovation processes, also comparing emerging innovation systems, such as genomics, with more stabilised ones, such as IT.

References

Akrich, M (1995). User representations: practices, methods and sociology. In *Managing Technology in Society. The Approach of Constructive Technology Assessment*, A Rip, J Misa and J Schot (eds.), pp. 167–184. London and New York: Pinter Publishers.

Atun, RA, I Gurol-Urganci and D Sheridan (2007). Uptake and diffusion of pharmaceutical innovations in health systems. *International Journal of Innovation Management*, 11(2), 299–321.

Atun, RA and D Sheridan (2007). Editorial: innovation in health care: the engine of technological advances. *International Journal of Innovation Management*, 11(2), 5–10.

Autio, E, S Kanninen and R Gustafsson (2008). First- and second-order additionality and learning outcomes in collaborative R&D programs. *Research Policy*, 37(1), 59–76.

Bijker, WE (1995). *Of Bicycles, Bakelites, and Bulbs. Toward a Theory of Sociotechnical Change*. Cambridge, MA: The MIT Press.

Boon, WPC, EHM Moors, S Kuhlmann and REHM Smits (2008). Demand articulation in intermediary organisations: the case of orphan drugs in the Netherlands. *Technological Forecasting and Social Change*, 75(5), 644–671.

Boon, WPC (2007a). *Storyline of Herceptin Reimbursement — Case Study Report on the Dutch Breast Cancer Association*. Utrecht: Utrecht University.

Boon, WPC (2007b). *Case Study Report on the Dutch Neuromuscular Disease Association*. Utrecht: Utrecht University.

Boon, WPC (2008, In Press). *Demanding Dynamics — Demand Articulation of Intermediary Organisations in Emerging Pharmaceutical Innovations* (dissertation). Utrecht: Utrecht University.

Boschma, RA and JG Lambooy (1999). Evolutionary economics and economic geography. *Journal of Evolutionary Economics*, 9(4), 411–429.

Boschma, R (2005). Proximity and innovation: a critical assessment. *Regional Studies* — Abingdon 00039/00001 (2005-01-01), 61–75.

Bucchi, M and F Neresini (2008). Science and public participation. In *The Handbook of Science and Technology Studies*, EJ Hackett, O Amsterdamska, M Lynch and J Wajcman (eds.), pp. 449–472. Cambridge, Massachusetts: The MIT Press.

Clark, KB (1985). The interaction of design hierarchies and market concepts in technological evolution. *Research Policy*, 14, 235–251.

Cohen, WM and DA Levinthal (1990). Absorptive capacity: a new perspective on learning and innovation. *Administrative Science Quarterly*, 35(1), 128–152.

Collingridge, D (1980). *The Social Control of Technology*. London: Pinter.

COM (2007). Together for health: a strategic approach for the EU 2008–2013. *Proc. of the White Paper, Presented by the Commission of the European Communities*, Brussels, http://ec.europa.eu/health/ph_overview/strategy/health_strategy_en.htm [23 October 2007].

Coombs, R et al. (2001). *Technology and the Market. Demand, Users and Innovation*. Northampton, MA: Edward Elgar.

Cooper, RG (1993). *Winning at New Products: Accelerating the Process from Idea to Launch*, 2nd Ed., Cambridge, MA: Addison-Wesley.

Dasgupta, P and PA David (1994). Toward a new economics of science. *Research Policy*, 23, 487–521.

DNC (Dutch Nutrigenomics Consortium), internal document for website www.nutrigenomicsconsortium.nl.

Doloreux, D (2004). Regional networks of small and medium sized enterprises: evidence from the metropolitan area of Ottawa in Canada. *European Planning Studies*, 12(2), 173–189.

Dosi, G et al. (2006). How much should society fuel the greed of innovators?: on the relations between appropriability, opportunities and rates of innovation. *Research Policy*, 35(8), 1110–1121.

FDA (2005). Innovation or stagnation. Challenge and opportunity on the critical path to new medical products. FDA report.

Fischoff, B (1995). Risk perception and communication unplugged: twenty years of progress. *Risk Analysis*, 15(2), 137–145.

Fleck, J (1994). Learning by trying. The implementation of configurational technology. *Research Policy*, 23(6), 637–652.

Fleck, J (1988). Innofusion or diffusation? The nature of technological developments in robotics. Edinburgh PICT Working Paper.

Franke, N and E Von Hippel (2003). Satisfying heterogeneous user needs via innovation toolkits: the case of Apache security software. *Research Policy*, 32(7), 1199–215.

Feldman, MP (1994). *The Geography of Innovation*. Dordrecht: Kluwer.

Frenken, K and FG Van Oort (2004). The geography of research collaboration: theoretical considerations and stylised facts in biotechnology in Europe and the United States.

In *Regional Economies as Knowledge Laboratories*, P Cooke and A Piccaluga (eds.), pp. 38–57. Cheltenham, UK and Northampton, MA: Edward Elgar.

Garrety, K and R Badham (2004). User-centered design and the normative politics of technology. *Science, Technology & Human Values*, 29(2), 191–212.

Geels, FW (2002). Understanding the Dynamics of Technological Transitions. A co-evolutionary and socio-technical analysis. PhD thesis, Twente University Press.

Gertler, MS (2003). Tacit knowledge and the economic geography of context, or the undefinable tacitness of being (there). *Journal of Economic Geography*, 3, 75–99.

Geurts, J (1993). *Omkijken naar de toekomst, lange termijn verkenningen in beleidsexercities (Looking Back to the Future, Longterm Foresight Studies in Policy Exercises)*. Tilburg: Samson H.D. Tjeenk Willink.

Griffin, A and JR Hauser (1993). The voice of the customer. *Marketing Science*, 12(1), 1–27.

Griffin, A (1996). Obtaining customer needs for product development. In *The PDMA Handbook of New Product Development*, MD Rosenau, Jr. (ed.), pp. 153–166. New York: John Wiley & Sons Inc.

Hamel, G and CK Prahalad (1994). *Competing for the Future*. Boston, MA: Harvard Business School Press.

Hoogma, R and JW Schot (2001). How innovative are users? A critique of learning-by-doing and -using. In *Technology and the Market. Demands, Users and Innovation*, R Coombs, K Green, A Richards and V Walsh (eds.), pp. 216–233. Cheltanhem, UK, Edward Elgar.

Howells, JRL (2002). Tacit knowledge, innovation and economic geography. *Urban Studies*, 39(5), 871–884.

Kaput, J, JM Ordovas, L Ferguson, B van Ommen, RL Rodriguez and L Allen (2005). Horizons in nutritional science. The case for strategic alliances to harness nutritional genomics for public and personal health. *British Journal of Nutrition*, 94, 623–632.

Kaput, J, A Perlina, B Hatipoglu, A Bartholomew and Y Nikolsky (2007). Nutrigenomics: concepts and applications to pharmacogenomics and clinical medicine. *Pharmacogenomics*, 8(4), 369–390.

Lettl, C, C Herstatt and HG Gemuenden (2006). Users' contribution to radical innovation: evidence from four cases in the field of medical equipment technology. *R&D Management*, 36(3), 251–272.

Lundvall, BA (1988). Innovation as an interactive process: From user-producer interaction to the national system of innovation. In *Technical Change and Economic Theory*, Dosi, G, C Freeman, R Nelson, G Silverberg and L Soete (eds.), pp. 349–369. London: Pinter Publishers.

Lundvall, BA (1992). *National Systems of Innovation: Towards a Theory of Innovation and Interactive Learning*. London: Pinter Publishers.

Lütje, C (2003). Customers as co-inventors: an empirical analysis of the antecedents of customer-driven innovations in the field of medical equipment. *Proc. of the 32nd Annual Conference of the European Marketing Academy (EMAC)*, Glasgow.

Malmberg A and P Maskell (1999). Guest editorial: localized learning and regional economic development. *European Urban and Regional Studies*, 6(1), 5–8.

Martin, P (2001). Great expectations: the construction of markets, products and user needs during the early development of gene therapy in the USA. In *Technology and the Market. Demand, Users and Innovation*, R Coombs, K Green, A Richards and V Walsh (eds.), pp. 38–67. Northampton, MA: Edward Elgar.

Menrad, K (2003). Market and marketing of functional food in Europe. *Journal of Food Engineering*, 56, 181–188.

Moors, EHM, C Enzing, A Van der Giessen and REHM Smits (2003). User-producer interactions in functional genomics innovations. *Innovation: Management, Policy & Practice*, 5(2–3), 120–143.

Moors, EHM and H Schellekens (2008). Medical biotechnology and sustainable drug development. In *General Aspects in Biopharmaceuticals for European Hospital Pharmacists*, H Schellekens and AG Vulto (eds.), Chap. 1, Mol(B): Pharma Publishing & Media Europe 2008, 7.

Nahuis, R and H Van Lente (2008). Where are the politics? Perspectives on democracy and technology. *Science, Technology & Human Values* 33(5).

Nahuis, R, EHM Moors and REHM Smits (2008). User producer interaction in context. A typology. Working Paper. *Innovation Studies Group*, Utrecht: Utrecht University.

Nelson, RR and SG Winter (1982). *An Evolutionary Theory of Economic Change*. Cambridge/London: The Belknap Press of Harvard University Press.

Oudshoorn, N and T Pinch (2003). *How Users Matter. The Co-construction of Users and Technology*. Cambridge, MA: MIT Press.

Poole, MS, AH Van de Ven, K Dooley and ME Holmes (2000). *Organizational Change and Innovation Processes*. Oxford: Oxford University Press.

Renn, O (1998). Three decades of risk research: accomplishments and new challenges. *Journal of Risk Research*, 1(1), 49–72.

Rip, A and R Kemp (1998). Technological change. In *Human Choice and Climate Change*, S Rayner and EL Malone (eds.), Vol. 2, Chap. 6, pp. 327–399. Ohio: Battelle Press.

Rip, A and JW Schot (2002). Identifying loci for influencing the dynamics of technological development. In *Shaping Technology, Guiding Policy*. KH Sørensen and R Williams, (eds.), pp. 155–172. Cheltenham, UK: Edward Elgar.

Rohracher, H (2005). *User Involvement in Innovation Processes. Strategies and Limitations from a Socio-Technical Perspective*. Munchen: Profil Verlag.

Rosenberg, N (1982). *Inside the Black Box. Technology and Economics*. Cambridge, MA: Cambridge University Press.

Royal Society (2005). *Personalised Medicine: Hopes and Realities*. London: The Royal Society.

Sabatier, PA (1987). Knowledge, policy-oriented learning and policy change: an advocacy coalitions framework. *Knowledge*, 8, 17–50.

Schmidt, JB and RJ Calantone (2002). Escalation of commitment during new product development. *Journal of the Academy of Marketing Science*, 30, 103–118.

Silverstone, R and E Hirsch (1992). *Consuming Technologies. Media and Information in Domestic Spaces*. London and New York: Routledge.

Smits, REHM (2002). Innovation studies in the 21th century: questions from a user's perspective. *Technological Forecasting and Social Change*, 69, 861–883.

Smits, R and P Den Hertog (2007). Technology assessment and the management of technology in economy and society. *International Journal of Foresight and Innovation Policy*, 3(1), 28–52.

Smits, REHM and WPC Boon (2008). The role of users in innovation in the pharmaceutical industry. *Drug Discovery Today*, 13(7/8), 353–359.

Star, SL and JR Griesemer (1989). Institutional ecology, 'translations' and boundary objects: amateurs and professionals in Berkeley's Museum of vertebrate zoology, 1907-39. *Social Studies of Science*, 19, 387–420.

Stewart, J and R Williams (2005). The wrong trousers? Beyond the design fallacy: social learning and the user. In *Handbook of Critical Information Systems Research: Theory and Application*, D Howcroft and E Trauth (eds.), pp. 195–221. Cheltenham: Edward Elgar.

Stirling, A (1998). Risk at a turning point? *Journal of Risk Research*, 1(2), 97–110.

Teubal, M (1979). On user needs and need determination: aspects of the theory of technological innovation. In *Industrial Innovation. Technology, Policy and Diffusion*, MJ Maker (ed.), pp. 266–289. London: Macmillan Press.

Triggle, D (2007). Treating desires not diseases: a pill for every ill and an ill for every pill? *Drug Discovery Today*, 12(3/4), 161–166.

Trusheim, MR, ER Berndt and FL Douglas (2007). Stratified medicine: strategic and economic implications of combining drugs and clinical biomarkers. *Nature Reviews Drug Discovery*, 6, 287–293.

Urala, N and L Lähteenmäki (2007). Consumers' changing attitudes towards functional food. *Food Quality and Preference*, 18, 1–12.

Utterback, JM (1994). *Mastering the Dynamics of Innovation*. Boston, Massachussets: Harvard Business School Press.

Vandeberg, RLJ and EHM Moors (2007). A framework for interactive learning in emerging technologies. Working Paper. *Innovation Studies Group*. Utrecht: Utrecht University.

Vandeberg, RLJ and WPC Boon (2008). Anticipating emerging technologies. *Genomics, Society, Policy Journal*. (Forthcoming).

Vandeberg, RLJ (2008a). *Interactive Learning in Emerging Technologies: the Case of the Dutch Nutrigenomics Consortium* 2007. Utrecht: Utrecht University — Department of Innovation Studies.

Vandeberg, RLJ (2008b). *Interactive Learning in Emerging Technologies: the Case of the German Competence Network Metabolic Syndrome*. 2007. Utrecht: Utrecht University — Department of Innovation Studies.

Vandeberg, RLJ (forthcoming). A framework for interactive learning in emerging technologies — Comparative case studies in national nutrigenomics consortia. Utrecht: Utrecht University (dissertation).

Van der Valk, T (2007). *Technology Dynamics, Network Dynamics and Partnering — The Case Of Dutch Dedicated Life Sciences Firms*. Utrecht: Utrecht University.

Van Kleef, E, HCM Van Trijp and P Luning (2005). Consumer research in the early stages of new product development: a critical review of methods and techniques. *Food Quality and Preference*, 16(3), 181–201.

Van Lente, H, MP Hekkert, REHM Smits and B Van Waveren (2003). Roles of systemic intermediaries in transition processes. *International Journal of Innovation Management*, 7(3), 1–33.

Verbeke, W (2005). Consumer acceptance of functional foods: sociodemographic, cognitive and attitudinal determinants. *Food Quality and Preference*, 16, 45–57.

Von Hippel, E (1988). *The Sources of Innovation*. New York: Oxford University Press.

Weterings, ABR (2006). *Do Firms Benefit From Spatial Proximity?* Utrecht: Utrecht University.

Part III
New Directions in User Innovation Research and Policy

OUTLAW COMMUNITY INNOVATIONS

CELINE SCHULZ and STEFAN WAGNER

University of Munich

Introduction

User innovation is a commonly observed phenomenon in various industries, where users modify or improve products that they use (von Hippel, 1988, 2005). Communities of user innovators — innovation communities — provide platforms for users, to openly and voluntarily communicate with each other regarding innovations they are working on either collectively or independently (Franke and Shah, 2003; von Hippel, 2005; Hienerth, 2006). Recent studies that have examined the relationship between innovation communities and firms have found that there is often a symbiotic relationship from which both users and manufacturers benefit (Jeppesen and Molin, 2003; Jeppesen and Frederiksen, 2006; Prügl and Schreier, 2006). Dahlander and Magnusson (2005) further distinguish between commensalistic (where the manufacturer gains and the community is indifferent) and parasitic (where the manufacturer gains on the expense of the community) relationships.

However, there could be an additional type of relationship where community innovations can be beneficial for users, and at the same time harmful for manufacturers. An example of such a relationship is when innovations stemming from communities, aim at bypassing legal or technical safeguards that prevent users from unsolicited usage of the manufacturer's products (Mollick, 2004). In particular, manufacturers of consumer electronic devices often install security mechanisms to prevent users from executing unauthorised software code or digital rights management (DRM)-protected content on their platform (such as pirated copies of authorised software or illegal copies of MP3s). Research on user innovations that deactivate such security mechanisms, in order to give the user full control over the usage of the product, was first introduced by Mollick (2004). Extending this research, Flowers (2008) coined the term *outlaw innovation* and provided case studies of how communities create and disseminate innovations that not only conflict with manufacturers' intensions of the usage of the original product but also violate firms' intellectual property rights (IPR). He proposes that outlaw communities consist of both users who innovate and those who simply adopt and use outlaw innovations. Recent examples of such outlaw communities in the consumer electronics industry include www.XBox-scene.com, www.XBox-linux.org and www.free60.org for the Microsoft XBox (a gaming console); www.iphonehacks.com for Apple's iPhone (a mobile phone); and www.cellphonehacks.com for cell phones in general. In the examples listed above, user innovators were able to "hack" (disable) security mechanisms enabling them (and other users) to run both user written software code (homebrew) and pirated programs on the now unprotected hardware.

Although both Mollick (2004) and Flowers (2008) perceive outlaw innovation as potentially harmful for the manufacturer, both propose strategies as to how firms can respond to these outlaw innovations in such a way that they can be profitable for their own innovation processes. But before firms can leverage outlaw communities for innovation management, they need to have a better comprehension of the phenomenon of outlaw innovation so that they can relate more effectively to it. It is thus important to first understand more about the actual individuals who modify and hack their products. To address these issues, our study aims to investigate the characteristics of the outlaw innovations implemented by users and the motivations that drive users to develop and adopt outlaw innovations.

In our paper, we provide results from an online-survey of users of two outlaw communities focusing on Microsoft's XBox. In total, we received 2256 questionnaires from our online-survey posted at www.XBox-scene.com and www.XBox-linux.org. About 20% of the responses originated from users who actively contributed to modifications of original software or to self-written (so called homebrew) software. Results indicate that users modify their XBox mainly to be able to increase the set of available functions of their XBox. Next, users are motivated to modify their XBox for the sake of having fun and to be able to run pirated games. We also find that user innovators are largely intrinsically motivated by fun and the intellectual stimulation of writing code for homebrew software.

The remainder of this paper is presented as follows: in the section "Outlaw Communities" we provide an overview of relevant literature on user communities and innovation; in the section "Survey Design" we have introduced the object of study — Microsoft's XBox and online-based outlaw communities focussing on the XBox's modification; the results of our survey are presented in the section "Empirical Results"; in the "Conclusion" section we conclude with a summary of the major findings and provides a short outlook on future research topics.

Outlaw Communities

User communities can be defined as horizontal user networks that consist of user nodes inter-connected by information transfer links which may involve face-to-face, electronic or any other form of communication that provide members sociability, support, a sense of belonging and social identity (Wellman et al., 2002; von Hippel, 2007). Three conditions are necessary for user communities to function entirely independently of manufacturers: (1) some users innovate, (2) some users freely reveal[1] their innovations and (3) users can self-manufacture their innovations relatively cheaply (von Hippel, 2007).

Communities can either focus on a common interest or on one or more products of a particular manufacturer. The first refers to special interest communities, such as sports communities where members interact with each other to exchange information and to assist each other with the development of their sports-related consumer product innovations (Franke and Shah, 2003; Shah, 2005a; Hienerth, 2006). The second type of communities consist of individuals, who all use a similar product and who learn how to use it better as they interact with others on a regular basis (Wenger, 2004). Examples of such user communities that have been examined in the literature include a firm-established user community for computer-controlled music instruments (Jeppesen and Frederiksen, 2006), user communities in the video gaming industry (Prügl and Schreier, 2006), a statistical software community (Mayrhofer, 2005) and user communities of a proprietary software firm (Schulz, 2006).

In the above-mentioned cases for communities that focus on products of a particular manufacturer, it was found that both the participants of the communities as well as the manufacturers of the underlying product profit from the existence of organised innovation communities. In particular, participants not only benefit from the user-to-user support but also from the gain in reputation they receive from the firm and other community members. When user innovators in these communities freely reveal their innovations, firms profit by implementing these innovations into their proprietary products and selling them to all users (Mayrhofer, 2005; Jeppesen and Frederiksen, 2006).

However, user communities may not always be beneficial for the manufacturer of the underlying product *per se*. Our study focuses on communities that exist independently from the manufacturer and whose members create innovations that aim at bypassing legal or technical safeguards that prevent users from unsolicited usage of the manufacturer's product (Mollick, 2004). Specifically, we examine one particular form of user community innovation — hacking, which refers to the user modifications of a product to not only gain unauthorised access to the product's system but also to enable the user to use the system more effectively (Meyer, 1989). According to Mollick (2004), hackers can be categorised into innovative "Elites" who understand the proprietary system and cause it to do things its makers never

---

[1] "Free revealing" is defined as the granting of access to all interested agents without imposing any direct payments.

intended, and follower "Kiddies" who do not truly understand the system, but merely use tools created by the Elites in order to exploit the system in their own way. Flowers (2008) generalises these concepts and defines outlaw communities, as consisting of both the users who are user innovators (Elites) and adopters of these user innovations (Kiddies). Specifically, outlaw communities can be defined as groups of users, who create and disseminate innovations that not only conflict with manufacturers' intentions of the usage of the original product but also violate firms' IPR.

Software hackers initiated the open source movement in the late 1990s, to promote open collaborative projects between software developers (Raymond, 1999). Although open source software communities today are widely perceived to be legitimate and "aboveground" (Mollick, 2004), it is nevertheless useful to examine the motivations that drive software developers (user innovators) to contribute to open source software projects, as they may be relevant for analysing outlaw communities. Theoretical and empirical studies show that these incentives for contribution can be categorised into intrinsic and extrinsic motivations. Extrinsic motivation refers to the separable outcome (or indirect reward) that is attained when an activity is done, whereas intrinsic motivation is the inherent satisfaction of the doing of an activity (Ryan and Deci, 2000). Examples of extrinsic motivations include skill improvement through active peer review (Hars and Ou, 2001; Ghosh et al., 2002) and the signalling motive for career advancements and/or future career benefits (Lerner and Tirole, 2001, 2002). Creativity (Lakhani and Wolf, 2005), fun (Torvalds and Diamond, 2001; Bitzer et al., 2004; Lakhani and Wolf, 2005) reputation (Raymond, 1999; Ghosh et al., 2002; Lakhani and Wolf, 2005), altruism (Zeitlyn, 2003; Bitzer et al., 2004) and reciprocity (Lakhani and von Hippel, 2003) are examples of intrinsic motivations.

In his study of four outlaw communities, Mollick (2004) finds that user innovators are intrinsically motivated by the desire to discover and innovate, rather than for the sake of theft. Adopters, on the other hand, are more often "vandals" who are motivated to adopt an outlaw innovation to gain unauthorised access to their product's system for pirate behaviour. Despite these differences, the user innovators are willing to allow the adopters to use their innovation, so as to gain attention to their work and to give them a certain sense of satisfaction, i.e., they benefit from the gain in reputation. This symbiotic relationship between the innovators and the adopters promotes the diffusion of outlaw innovations.

Differentiating between user innovators and adopters, we aim to investigate the characteristics of outlaw innovations which are implemented by community members and the motivations that drive individuals to develop and adopt outlaw innovations. In particular, we hope to be able to provide a deeper understanding regarding the inherent nature of the outlaw innovation, that is, if it is intrinsically "harmful" or if they are just a result of creative individuals having fun.

Survey Design

Object of study — The case of Microsoft's Xbox

The phenomenon of outlaw community innovation has already been identified and described in the literature (Mollick, 2004; Flowers, 2008). However, until now it remains mostly unclear what exactly drives individuals to participate in such communities and to what extent the two different types of community participants (user innovators and adopters) described by Mollick (2004) and Flowers (2008) do co-exist in outlaw communities. While survey evidence of the innovation communities exists (Franke and Shah, 2003; Shah, 2005a; Hienerth, 2006; Jeppesen and Frederiksen, 2006; Prügl and Schreier, 2006), we present results from the first survey of participants of two large Internet-based outlaw innovation communities focused on Microsoft's XBox. Before advancing to the description of the survey design applied, we would like to provide some background information on the emergence of user modifications of various electronic products in general, and in particular modifications of the XBox.

First, it is important to note that the technical architecture of many consumer electronic products is based on the design of standard personal computers. They contain key building blocks like a central processing unit (CPU) which processes commands initiated by an (often embedded) operating system,[2] memory devices for manipulating and storing data and different types of input/output devices such as displays and controllers or keyboards. In general, the embedded operating system (often called firmware) allows users to execute a closed set of exactly specified functions on the hardware.[3] The set of defined functions provided by the firmware is only a subset of all functions which could in theory be performed by an electronic device. Sometimes, consumer electronic devices like video gaming consoles or mobile phones also allow the execution of third-party applications like games or other pieces of software. However, the execution of additional applications is often restricted by security mechanisms requiring third-party developers to acquire costly licenses from the device's manufacturer, to ensure compatibility with the embedded firmware.

Most user innovations occur in this context as attempts to manipulate or to replace the embedded operating system in order to extend the limited set of functions provided by the manufacturer, with additional functions or to disable security mechanisms to allow for the execution of unauthorised pieces of software as

---

[2] An operating system can be defined as *embedded* if it is contained inseparably in a device. Often embedded operating systems are stored on ROM or flash memory chips and cannot easily be replaced by the user.

[3] In the following, "firmware" specifies the operating system embedded in an electronic device by its manufacturer.

well as illegal copies of authorised software. The manipulations of the firmware which bypasses security mechanisms largely resembles the hacking of computer systems and is therefore often regarded as an outlaw activity (see Flowers (2008) for a more detailed discussion). Known examples of products where users provided modifications/hacks of the embedded operating system range from mobile phones (Apple's iPhone has been hacked in order to allow for the execution of unauthorised third party applications (See http://www.iphonehacks. com for more details. Latest visit on March 26th 2008)), to hard-disk based VCRs (See http://www.lugod.org/presentations/tivohacks for more details. Latest visit on March 26th 2008) and network controllers (See http://www.nslu2-linux.org/wiki/Unslung/HomePage for more details. Latest visit on March 26th 2008).

It should be noted, though that video gaming consoles and video games themselves have historically been a target of user modifications since their emergence (see Levy, 1984 for a detailed overview). In general, consoles are a particularly attractive target for user modifications for several reasons: First, gaming consoles are most often powerful computing devices which are often equipped with high performance CPUs and graphic processors in order to cope with computational intense high-definition graphics of video games. As a consequence, gaming consoles can, in principle, execute a wide range of different applications without running into computational restrictions. Second, gaming consoles are often sold at a subsidised price as manufacturers anticipate significant revenue streams from complementary products such as software, games or controllers (Soghoian, 2007). In many cases, purchases of hardware with similar technical specifications would be possible only at higher prices.

While user modifications exist for almost all major video gaming consoles (Grand et al., 2004), we will focus on Microsoft's XBox for a number of reasons.[4] When it was released in 2001, the XBox was equipped with a 733-MHz Intel Celeron processor, a nVidia GeForce GMX graphics processing unit running at 233 MHz, a 100 MB/second ethernet interface and a 10 GB hard disk (see Grand et al., (2004) for a detailed description of the technical specifications of the XBox). This configuration easily topped even upper-scale desktop computers sold at the same time.[5] Despite offering immense computational power, the initial recommended price of an XBox was relatively low at USD299, in the US as comparable to personal computers with a similar performance (See http://en.wikipedia.org/wiki/XBox. Latest

---

[4] We focus on the first-generation XBox. The current version of the XBox ("XBox 360") was introduced in December 2005.

[5] Due to the comparably long life cycles of gaming consoles — for example, the XBox was sold until 2005 with an unchanged hardware configuration — manufacturers include state-of-the-art technology when launching new products.

visit February 26th 2008). Due to the PC-like architecture and its comparably low price, the XBox quickly drew the attention of user innovators trying to disable Microsoft's security mechanism, which was eventually hacked at the end of 2001 (Takahashi, 2002). As a consequence, users were able to execute unauthorised software (including illegal copies of authorised software) on the Xbox, provided they had the appropriate hacks to disable Microsoft's security mechanism.[6]

After the XBox's security mechanism was hacked, numerous pieces of software provided by individual users emerged (so called *homebrew* software). As it would be far beyond the scope of this paper to provide a detailed overview on the XBox homebrew-scene as a whole, we restrict ourselves to the organisation of developers and adopters of homebrew software in Internet-based communities where information on the software is discussed and distributed. The development of homebrew software within these communities often follows the principles of open source software development where a larger group of developers collaborate on large software suites for which source code is put into the public domain. One of the most successful homebrew applications for Microsoft's XBox is the XBox media centre (XBMC) (see www.xbmc.org for a detailed description) — a very comprehensive multi-media application providing possibilities to store, administrate and play different types of multi-media files including audio and video files. Note that the original XBox firmware provided by Microsoft provides only very limited possibilities to exploit the multi-media skills of the XBox's hardware. It offers the possibility to play and store music from CDs inserted directly to the XBox with limited functionality with regard to the management of large collections of songs. The playback of DVDs is only possible if an additional controller is purchased by the user. The XBMC extends this limited set of multi-media skills of the XBox with various functions and can, therefore, be seen as users' efforts to increase the set of available functions. It is worth emphasising that the collaborative effort in this project yielded a multi-media package whose quality is above the standard of comparable players provided for PC operating systems — in fact, the XBMC is so successful that it has even been ported to the Apple OS.

Survey design

To date, studies on the phenomenon of outlaw innovation focused on qualitative case studies (Mollick, 2004; Flowers, 2008) and there is little systematic large scale evidence with respect to characteristics of users engaged in such activities or their motivations. In order to shed more light into these questions, we conducted an online

---

[6]The XBox's security mechanism can be disabled by overwriting an internal flash memory chip, by adding an additional chip or by applying a software-based hack.

survey of Internet communities focussing on homebrew software for the XBox. We surveyed participants of two relevant online communities using an anonymous online questionnaire hosted on servers of the University of Munich between March 15th and April 3rd, 2005. The questionnaire had been developed based on a series of interviews of active developers of XBox modifications including the webmasters of the online communities whose members we surveyed.[7] The first community we targeted was www.XBox-scene.com. According to this website:

> "XBox-scene's primary goal is to keep its visitors up-to-date about the XBox Scene. Unlike other sites, software pre/ reviews aren't [...] major sections. XBox-Scene is specialized in hardware news and information. [XBox-scene] will talk about mods (modifications) and other hardware aspects." (See http://www.XBox-scene.com/about.php for more details, latest visit on February 26th, 2008.)

The second online community we surveyed was the www.XBox-linux.org. While XBox-scene.com is a general interest community with an emphasis on the whole spectrum of possible modifications of the XBox, XBox-linux.org is closely related to the XBox Linux project which,

> "[...] aims to provide a version of GNU/Linux for the XBox, so that it can be used as an ordinary computer. Linux should make use of all XBox hardware and allow the user to install and run software from standard i386 Linux distributions". (See http://www.XBox-linux.org/wiki/XBox-Linux:About. Latest visit on February 26th, 2008.)

Our survey was advertised prominently on both web pages and resulted in a total of 2256 responses. Since the data on the size of the basic population was not available, we are unable to report a precise response rate. However, we conducted a non-response analysis comparing early to late respondents (Armstrong and Overton, 1977) which yielded no indication of a non-response bias.[8]

In our study, outlaw community members are defined as members who have modified their XBox to circumvent the XBox's security mechanism so as to be able

---

[7] We conducted a total of six unstructured interviews of about 90 minutes, each prior to designing the questionnaire. Our interview partners include the administrators of www.xbox-linux.org, www.xbox-scene.com, the designer of the SmarttXX modchip as well as three users owning a modified XBox.

[8] In particular, we compared the first 10% with the last 10% of the responses to our survey with regard to the means of relevant variables using t-tests. The observed differences are not significant and does not point to systematic differences between early and late responses.

to execute software, which is not authorised by Microsoft (homebrew software and illegal copies of authorised software). Based on the findings of Mollick (2004) that outlaw communities often consist of innovators and adopters, we aim to identify both groups of participants. We therefore asked our respondents to indicate whether they have actively contributed to the development of any unauthorised software code for the XBox. In total, approximately 20% of the respondents are user innovators who contributed to unauthorised code while 80% did not contribute any modifications to the community and can be classified as adopters accordingly.

All respondents to our survey are male and aged between 14 and 61 years. The average age of our respondents is 23 years. Although most of the respondents have a college degree or higher (43%), about a fifth of them are still in school and approximately a third of them have only a high school degree. The majority of the respondents came from the North America (U.S. 47.9% of all respondents and Canada 9.7% of all respondents). The remaining respondents are distributed more or less equally across Europe and the Asia-Pacific region.

Empirical Results

In the following two sub-sections, we present results of our survey which describe firstly the process of modifying the XBox and secondly the motivation to do so for all respondents of our survey. If appropriate, we will contrast the answers of user innovators and adopters in the surveyed communities in order to highlight relevant differences. In the section "characteristic of user innovators", we present evidence of the major motivations of user innovators and compare them to previous evidence of user innovators which have been derived from surveys conducted within the open source software scene.

Modification and use of the Xbox

The disablement of the XBox's security mechanism is a pre-requisite for the execution of non-authorised software code. There are various ways to circumvent the security mechanism. All of them require some modifications of the gaming console — either by modification of the hardware configuration or by modification of the embedded firmware. One way of performing the necessary modification is to open up the XBox and to solder in a modchip which is a specifically customised chip disabling the security mechanism.[9] Alternatively, it is also possible to overwrite (flash) the XBox's ROM-chip containing the embedded firmware with alternative software (hardware hack). A third way is to exploit a buffer over-run, by applying

---

[9]Modchips for video gaming consoles are legally sold by third party manufacturers.

Table 1. Frequency table of how respondents modified their XBoxes. Percentages in brackets.

Who performed the modification of your XBox	Adopters	Innovators	Total
I bought a modified Xbox (new)	67	8	75
	3.73%	1.96%	3.40%
I bought a modified Xbox (used)	20	3	23
	1.11%	0.73%	1.04%
I did it my own (without any help)	1,224	322	1546
	68.23%	78.73%	70.18%
I did it with help from my friends	262	42	304
	14.60%	10.27%	13.80%
I shipped the Xbox to a commercial vendor	88	13	101
	4.91%	3.18%	4.58%
Others	78	16	94
	4.35%	3.91%	4.27%
Assistance from the Xbox-Linux Community	55	5	60
	3.07%	1.22%	2.72%
Total	1,794	409	2203

special software tools (software hack). In the latter two cases, it is not necessary to open up the XBox. According to our survey, approximately three quarters of the respondents (78.6%) had soldered a modchip into their XBox, 23% had overwritten the ROM-chip and 25% performed a software hack — multiple hacks on the same Xbox are possible.

Table 1 presents the results of how individuals modified their XBox: 70.3% of the respondents indicated that they had modified their XBox themselves while 13.7% were assisted by friends. Results differ slightly across innovators and adopters. A larger percentage of user innovators (78.7%) modified their XBox themselves as compared to the adopters (68.2%). In a t-test, we were able to reject the hypothesis that this difference is equal to zero at the 1% significance level.

Once the XBox has been modified in a way that the security mechanism is disabled different types of unauthorised software codes can be executed. Figure 1 shows the percentage of respondents, who installed various types of software applications. In addition to the extension of the XBox's spectrum of functions by the installation of software, hardware modifications can be performed to improve the technical specifications of the console. For example, the XBox contains a standard IDE hard drive controller making it easy for users to exchange the original 10 GB hard disk by a larger standard 3.5″ hard disk in order to increase storage capacity. Figure 1 shows that almost all participants had installed the XBox Media Center XBMC (94.6% of all respondents), and amended the operating system (also called dashboard, 95.4% of all respondents). In combination with the large fraction of

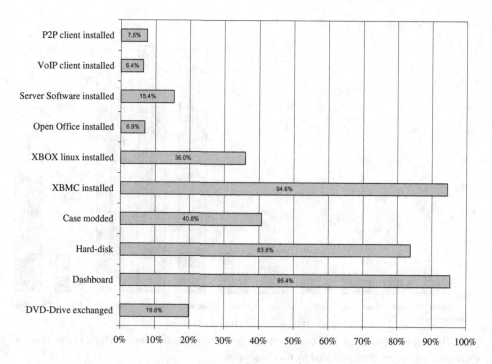

Fig. 1. Share of respondents who applied the following modifications to their Xbox.

respondents that exchanged the original hard disk (83.8%), it can be assumed that most of the respondents are using the XBox as a multi-media centre which enables the user to be able to store, manage and play different types of media files. This inference is supported from our interview evidence with active members of XBox user communities suggesting that the XBox with its direct TV-output is an ideal TV set-top box as it is not only able to play from but also to rip DVDs and CDs directly to its hard-disk. Note that the installation of software applications familiar from PCs (Linux, office applications, voice-over-IP-clients (VoIP) or also peer-to-peer-clients) do exist but are far less often installed than the ubiquitous XBMC.

We further investigated the usage of modified XBoxes as we asked users to indicate how often they use their XBox for certain purposes on a five-item scale (never, rarely, occasionally, often, always). Figure 2 signifies the frequency of respondents' usage of their Xbox separately for innovators and adopters. It can easily be seen that the large majority of the participants use the XBox primarily to play games, watch movies and listen to music. This result corresponds to the earlier finding that almost all participants had installed the XBox Multimedia Centre which turns the XBox into a multi-media device enabling users to watch movies and listen to music — functions which are provided by Microsoft's original firmware only to a very limited extent. Interestingly, user innovators tend to use their XBox as a

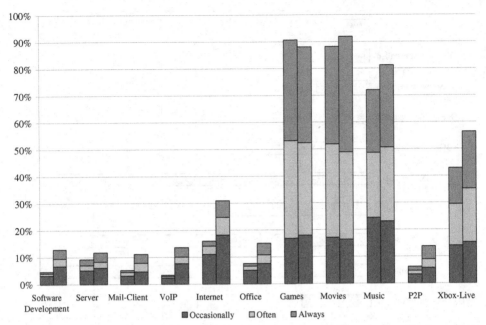

Note: The first bar of each category corresponds to the answers provided by adopters. The second bar corresponds to the answers provided by user innovators.

Fig. 2. Frequency of alternative ways to use the XBox.

regular computer more often than adopters. They use the modified console more frequently for software development, as an Internet-server or as a mail-client, for telecommunication purposes (VoIP), to surf the Internet, to run Microsoft Office applications and for file sharing over the Internet (P2P).

Motivation to modify the Xbox

In the sub-section "Modification and use of the XBox", we described the process of modifying and the use of the modified XBox without taking reference to the motivation of the users to engage in this behaviour. Motivational aspects deserve special attention here as the modification of the XBox often involves the breach of IPR[10] and can therefore be regarded as an "outlaw" activity.

As indicated in Fig. 3, one of the main motivations for users to modify their XBox is for the fun of hacking computer systems. Specifically, 63.4% of the adopters and 78.1% of the innovators indicated that this is an important to very important reason for modifying their XBox. We were able to reject the hypothesis that the difference

---

[10] Some of the tools enabling users to hack their XBox by modifying the embedded firmware violate Microsoft's IP rights.

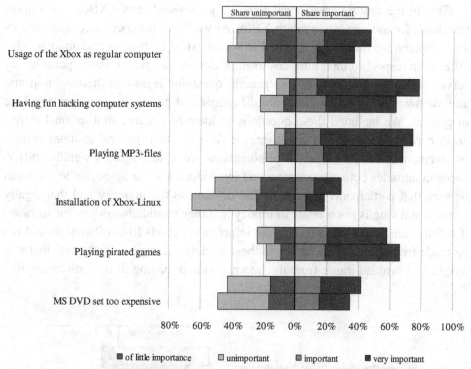

Fig. 3. Importance of different motivational aspects to modify the XBox.

Note: The upper bar of each category corresponds to the answers provided by innovators. The lower bar of each category corresponds to the answers provided by adopters. The neutral category "neither important nor unimportant" is not reported in the figure.

between the two groups is equal to zero at the 1% level of significance from the t-test. Another important motivation for hacking the console is to be able to execute illegal copies of games on the XBox. In particular, this aspect was rated as being important or very important by 66.1% of the adopters and 57.8% of user innovators.[11] Results from the statistical t-test suggest that the difference between adopters and user innovators is different from zero at the 1% significance level. This latter finding support the findings of Mollick (2004), who found that user innovators tend to be more intrinsically motivated to break into secured systems and argues that adopters tend to be "vandals", who are more motivated to adopt an outlaw innovation to gain unauthorised access to their product's system for deviant behaviour such as pirating authorised software.

---

[11] Additionally, we asked the respondents to indicate how many games they owned in total and how many of those are original versions. On an average, participants own approximately 53 games and only 13 are original.

Clearly, the ability to run illegal copies of games sold for the XBox is an important driver for users to modify their XBox. As we find differences across innovators and adopters, we further investigated to which extent innovators and adopters also differ with respect to their attitudes towards deviant behaviour. Participants of our survey were asked to answer five general questions regarding their opinion and attitude towards copyright protection of immaterial goods like software or music in general. We included these questions to identify whether motivational differences between adopters and innovators are driven by their general attitudes or their involvement in the development of homebrew software. Figure 4 signifies differences in attitudes between adopters and innovators. On the aggregate level, it can be seen that participants are not all that deviant, as the majority feel that legally attained software is not a waste of money that they would always pay for software if it fully satisfies their needs and the information goods like software should not be made free of charge. Contrary to these findings, a large majority feel that it is alright to download music from the Internet without paying. It is worth noting that

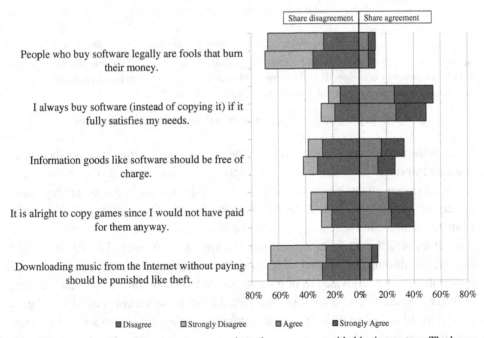

Note: The upper bar of each category corresponds to the answers provided by innovators. The lower bar of each category corresponds to the answers provided by adopters. The neutral category "neither agree nor disagree" is not reported in the figure.

Fig. 4. Agreement to different statements related to respondents' attitude towards copyright protection of immaterial goods.

user innovators do not necessarily have a greater tendency to respect copyrights — in fact, the differences of the distribution of the answers within the categories are not significantly different at the 5% level according to a χ^2-test. We interpret this finding as an indication that the observed lower importance of the possibility to run pirated software among innovators compared to adopters (for example, see Fig. 3) is not driven by innovators' attitude towards copyrights in general.

Characteristics of user innovators

So far, we reported results from our survey for all respondents distinguishing between innovators and adopters. As we are particularly interested in the motivation of user innovators to develop "outlaw" innovations, we now turn to their characteristics. All statistics reported in the following refer exclusively to the responses given by the 447 innovators (about 19.99% of all respondents) in our sample.

With regard to the number of projects, we observed that user innovators were involved in an average 2.67 different projects when answering our survey. Figure 5 indicates the share of innovators involved in the different types of software applications. It can be observed that the development of alternative operating systems (dashboard, 46%) and multi-media applications (46%) are the most frequent categories

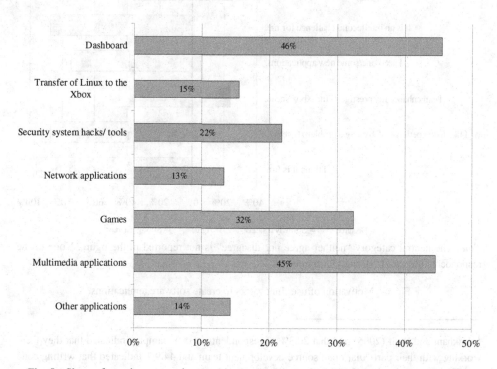

Fig. 5. Share of user innovators involved in the development of different types of applications.

for which user innovators develop software. This is followed by the development of add-ins for authorised games (32%) and software hacks and tools to gain unauthorised access to the XBox's system (22%). These findings are especially interesting, as they provide evidence that users are not only just trying to gain unauthorised access to the XBox's system but that the majority of them are creating innovations to be able to use the system more effectively.

To better understand why user innovators are motivated to create applications, respondents were further asked to indicate the major reasons for their innovative activities. Figure 6 demonstrates that almost all user innovators (96%) indicated that fun is a major incentive and 86% answered that they feel intellectually challenged when they create software applications for their XBox. Although these findings are much larger than those for open source software developers,[12] they

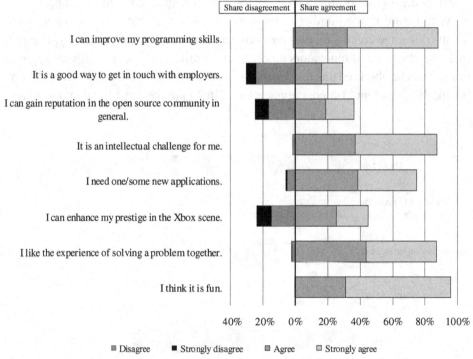

Note: The neutral category "neither agree nor disagree" is not reported in the figure. None of the respondents indicated "strongly disagree".

Fig. 6. Motivation of user innovators to create software applications.

---

[12]Lakhani and Wolf (2005) find that 20.3% of respondents in their sample, indicated that they liked working with their particular open source development team and 44.9% indicated that writing code for the open source project was intellectually stimulating.

support those of Mollick (2004) who found that user innovators tend to be intrinsically motivated by the desire to discover and innovate. Interestingly, we also find that a large majority of the innovators like the experience of solving problems together (87%), suggesting that user innovators benefit by communicating with each other on innovation projects they are working on. This finding is consistent with what Shah (2005b) found that by working together with other users, user-innovators were able to access additional resources to develop and improve their innovations.

It can also be inferred from Fig. 6 that extrinsic aspects are important for user motivation to innovate. In fact, 88% of the responding innovators indicate that the possibility to improve their programming skills motivates them to engage in the development of homebrew software. Further, innovators create applications because they need the applications for their own personal use. This result corresponds to the motivation proposed by von Hippel (1988) as to why users innovate. However, in contrast to findings of Mollick (2004), we find that the majority of the user innovators are not motivated by reputation. Specifically, only 45% indicated that they innovate to enhance their prestige in the XBox scene and even less (36%) indicated that they innovate to gain reputation in the open source software community in general.

Conclusion

It is a common phenomenon that users develop innovations, which modify or improve products provided by a manufacturer. If these innovations extend the possible use beyond the use envisaged by the manufacturer they can be regarded as "outlaw" innovations (Flowers 2008). Our survey of participants of two online innovation communities built around outlaw innovations for Microsoft's XBox, yielded interesting insights regarding the reasons why users engage in outlaw behaviour. Our findings are two-fold: First, we found that most users modify/hack their console in order to gain access to functionalities not provided by the manufacturer. In particular, almost all of the 2256 respondents installed the XBox MC, which provides much better multi-media functions than Microsoft's original firmware. Second, users are largely motivated to modify their XBox for the possibility to run pirated software but also for intrinsic motivations such as fun of hacking computer systems. With regard to the characteristics of user innovators, we find they are more intrinsically motivated than open source software developers. In particular, fun and intellectual stimulation are major reasons as to why they participate in the development of homebrew software.

Our findings suggest that the existence of outlaw communities favours the diffusion of outlaw innovation, such as that of the XBox MC which has been adopted by a large number of community members. In this respect, manufacturers benefit from the existence of such communities as they provide complements otherwise not available to their proprietary product, thus increasing its attractiveness. On the other hand, one other major finding is that participants in outlaw communities are also interested in pirate behaviour. Outlaw communities provide them with necessary information and tools to engage in such a pirate behaviour, which manufacturers perceive to be harmful to them. Therefore, future research should try to weigh the manufacturer's costs and benefits from the existence of outlaw communities. Our survey provides a first explorative starting point by attempting to clearly delineate what these costs and benefits from outlaw communities are.

Acknowledgment

We greatly acknowledge helpful discussions with Dietmar Harhoff and Ulrich Lossen. Further, we would also like to thank our colleagues at INNO-tec and participants of the TIME seminar at the TU Munich and participants of the "Conference on IP and the creative industries" in London in 2008.

References

Armstrong, JS and TS Overton (1977). Estimating nonresponse bias in mail surveys. *Journal of Marketing Research*, 14, 396–402.

Bitzer, J, W Schrettl and PJH Schröder (2004). Intrinsic Motivation in Open Source Software Development, Nr. 2004/19. Diskussionsbeiträge des Fachbereichs Wissenschaft der Freien Universität Berlin.

Dahlander, L and MG Magnusson (2005). Relationships between open source software companies and communities: observations from Nordic firms. *Research Policy*, 34(4), 481–493.

Flowers, S (2008). Harnessing the hackers: the emergence and exploitation of outlaw innovation. Cops working paper, University of Brighton.

Flowers, S (2008). Harnessing the hackers: the emergence and exploitation of outlaw innovation. *Research Policy*, 37(2), 177–193.

Franke, N and S Shah (2003). How communities support innovative activities: an exploration of assistance and sharing among end-users. *Research Policy*, 2(1), 157–178.

Ghosh, RA, R Glott, B Krieger and G Robles (2002). Free/libre and open source software: survey and study. Technical report, University of Maastricht and Berlecon Research GmbH: International Institute of Infonomics.

Grand, J, A Yarusso, R Baer, M Brown and F Thornton (2004). *Game Console Hacking: Having Fun While Voiding Your Warranty*. Rockland, MA: Syngress Publishing.

Hars, A and S Ou (2001). Working for free? Motivations for participating in open-source projects. *International Journal of Electronic Commerce*, 6(3), 25–39.

Hienerth, C (2006). The commercialization of user innovations: the development of the rodeo kayak industry. *R&D Management*, 36(3), 273–294.

Jeppesen, LB and L Frederiksen (2006). Why do users contribute to firm-hosted user communities? The case of computer-controlled music instruments. *Organization Science*, 17(1), 45–63.

Jeppesen, LB and MJ Molin (2003). Consumers as co-developers: learning and innovation outside the firm. *Technology Analysis & Strategic Management*, 15(3), 363–383.

Lakhani, KR and E von Hippel (2003). How open source software works: "free" user-to-user assistance. *Research Policy*, 32(6), 923–943.

Lakhani, KR and RG Wolf (2005). Why hackers do what they do: understanding motivation and effort in free/open source software projects. In *Perspectives on Free and Open Source Software*, J Feller, B Fitzgerald, S Hissam and KR Lakhani (eds.), pp. 3–22. Cambridge–London: The MIT Press.

Lerner, J and J Tirole (2001). The open source movement: key research questions. *European Economic Review*, 45(4–6), 819–826.

Lerner, J and J Tirole (2002). Some simple economics of open source. *The Journal of Industrial Economics*, 50(2), 197–243.

Levy, S (1984). *Hackers: Heroes of the Computer Revolution*. New York: Anchor Press/Doubleday.

Mayrhofer, P (2005). Does commercial interest hinder innovation in user communities? — a theoretical and empirical study. Unpublished Master's thesis, University of Munich.

Meyer, GR (1989). The social organization of the computer underground. Unpublished Master's thesis, Illinois: Northern Illinois University.

Mollick, E (2004). Innovations from the underground: towards a theory of parasitic innovation. Unpublished Master's thesis, Cambridge, MA: MIT Library.

Prügl, R and M Schreier (2006). Learning from leading-edge customers at the sims: opening up the innovation process using toolkits. *R&D Management*, 36(3), 237–250.

Raymond, ES (1999). The cathedral and the bazaar. In *The Cathedral and the Bazaar: Musings on Linux and Open Source by an Accidental Revolutionary*, pp. 27–78. Cambridge: O'Reily & Associates, Inc.

Ryan, RM and EL Deci (2000). Intrinsic and extrinsic motivations: classic definitions and new directions. *Contemporary Educational Psychology*, 25(1), 54–67.

Schulz, C (2006). The secret to successful user communities: an empirical analysis of computer associates' user groups. Discussion Papers in Business Administration 2006–2009, University of Munich, Munich School of Management.

Shah, S (2005a). From innovation to firm & industry formation: "innovation communities" in the windsurfing, skateboarding, and snowboarding industries. Working Paper 05-0107, University of Illinois.

Shah, SK (2005b). Open beyond software. In *Open Sources 2.0: The Continuing Evolution*, D Cooper, C DiBona and M Stone, (eds.), pp. 339–360. Sebastopol, CA: O'Reilly Media.

Soghoian, C (2007). Caveat venditor: technologically protected subsidized goods and the customers who hack them. *Northwestern Journal of Technology and Intellectual Property*, 6, 46–72.

Takahashi, D (2002). *Opening the Xbox: Inside Microsoft's Plan to Unleash an Entertainment Revolution*. Roseville, CA: Prima Publishing.

Torvalds, L and D Diamond (2001). *Just for Fun: The Story of an Accidental Revolutionary*. New York: HarperBusiness.

von Hippel, E (1988). *The Sources of Innovation*. Oxford–New York: Oxford University Press.

von Hippel, E (2005). *Democratizing Innovation*. Cambridge–London: The MIT Press.

von Hippel, E (2007). Horizontal innovation networks — by and for users. *Industrial and Corporate Change*, 16(2), 1–23.

Wellman, B, J Boase and W Chen (2002). The networked nature of community on and off the internet. *IT & Society*, 1(1), 151–165.

Wenger, E (2004). Knowledge management as a doughnut: shaping your knowledge strategy through communities of practice. *Ivey Business Journal*, 68(3), 1–8.

Zeitlyn, D (2003). Gift economies in the development of open source software: anthropological reflections. *Research Policy*, 32(7), 1287–1291.

USER INNOVATION: THE DEVELOPING POLICY RESEARCH AGENDA

STEPHEN FLOWERS

University of Brighton, UK

Introduction

The policy understanding of the processes, participants and dynamics of innovation is changing. The narrow 'linear' model of innovation that has dominated innovation policy for many decades is being reappraised and a broader definition is emerging. As part of this reappraisal the role of users is attracting considerable attention and research groups around the world are starting to collect the detailed evidence required to inform policy options and guide responses. User innovation is beginning to be drawn into mainstream innovation policy. This is a significant shift in the understanding and framing of innovation and represents a new phase in user innovation research that is specifically designed for innovation policy. This research will draw heavily on the conceptual and theoretical insights of the large existing body of user innovation research, but will also be shaped by the particular requirements of the policy audience.

This chapter will explore the emergence of user innovation policy research and outline the emerging research agenda. By way of an introduction to this new field it will briefly explore the linear model of innovation and also examine the wider changes in innovation policy that are beginning to emerge. The importance of metrics and indicators in developing innovation policy will also be examined and the user innovation policy research underway in Denmark, Canada, The Netherlands, the EU and the UK will be outlined. Early results from projects that are seeking to measure user innovation activity will be reported and chapter will conclude by outlining the emerging user innovation policy research agenda.

The Changing Face of Innovation Policy

Innovation policy is largely concerned with economic growth and international competitiveness and, although it may also be seen as a way to solve other social and economic problems, its main focus is on the creation of economic wealth (Lundvall and Borras, 2005). The traditional view concerning the rate of innovation within a country is that it depends on large number of inter-related factors including the volume of industrial R&D, the IP regime, the amount of publically-funded basic research, the education and training of scientists and engineers, access to venture capital, and government innovation policies (e.g. Sainsbury, 2007). The conceptual model that been used for over 50 years to frame this 'traditional' approach to innovation and innovation policy is beginning to reappraised. This model, often referred to as the linear model of innovation, provides a simple structure in which innovation is the outcome of basic research, applied research and development, production and diffusion.

Basic Research — Applied research — Development — Production and Diffusion

This simple, 'linear', model of innovation offers an entirely supply-side perspective that has been (and continues to be) highly influential. The linear model has provided the basis for much of the national and international policy infrastructure that has been developed to measure, support and promote innovation. The implications of this are not insignificant. For example, the innovation metrics and indicators widely used by many countries around the world all currently conform to this linear model of innovation that privileges industrial R&D over other forms of innovation. Certain measures have become a major focus of policy, with a recent UK Government report on innovation noting that 'The two most commonly used measures of innovative performance are the quantity of industrial research and the volume of patenting' (Sainsbury, 2007, p9). In part this reflects the role of the OECD in formalising the measurement of scientific and technological activities in the Frascati and Oslo Manuals, but it is also the result of the dominance of the supply-side approach towards innovation. This focuses on a particular view of innovation and the innovation system and can be seen in the orientation of innovation metrics and indicators.[1]

Science, Technology and Innovation (STI) indicators often have a clear firm and market bias. For example, the Oslo manual deals with innovation at the level of the individual firm and focuses on 'innovation in business enterprise sector' (OECD, 2002, p9), but in the context of user innovation this is problematic on a number of levels. Partly this arises from the definition of the term 'user' which can be used to refer to both firms and individual consumers.[2] In addition, although there is extensive

---

[1] However, it should be noted that this chapter does not attempt to offer a comprehensive critique of STI metrics and indicators. A comprehensive account may be found in Godin, B. (2005) Measurement and statistics on science and technology: 1920 to the present, Routledge.

[2] In the context of a mass-produced consumer good the 'user' of that product is likely to be an individual, whereas a firm who acquires a specialist process technology designed for the business-business

user innovation within firms (e.g. Arundel, 2008), much firm-level process innovation is non-commercial in nature and is thus falls outside the OECD's definition. Similarly, the extensive user innovation activities observed amongst individual users and user communities are take place outside firms and tend to be non-commercial in nature. As a result huge swathes of innovative activity simply does not appear within official innovation metrics and indicators, and are thus largely invisible to policy.

In contrast, Research and Development (R&D) metrics provide a very visible insight into firm-level innovation activities. R&D is defined as the 'creative work undertaken on a systematic basis in order to increase the stock of knowledge, including knowledge of man, culture and society, and the use of this stock of knowledge to devise new applications' (OECD, 2002, p31). In practical terms the collection of R&D data focuses on systematic and organized activities that favours formal firm-level R&D and overlooks informal activity. However, it has long been recognized that official R&D statistics have inherent limitations in this respect (e.g. OECD, 1976) and that there are other forms of non-R&D innovation (e.g. Kline and Rosenberg, 1986). It is also understood that highly innovative firms may not perform R&D in the 'official' sense (e.g. Arundel et al., 2008). Given this background much of the activity associated with user innovation both at a firm level (e.g. innovation in firm processes and productive equipment) and at an individual consumer or community level (scientific instruments, extreme sports, computer game modding) is not reflected in official R&D statistics.

In the same way patents and patent indicators offer a very particular insight into the innovation process that is of little relevance outside the linear model. A patent is a government right granted to an individual or organisation conferring a right or title to profit from an invention and patents relate to inventions and are not, in themselves, indicators of innovation outputs (OECD, 1996, p59). However, although they provide insights that are essential for a deeper understanding of aspects the traditional 'linear' process model of innovation, they have many drawbacks (e.g. see OECD, 1994). Patents also have little to offer our understanding of user innovation. Research has shown that many users (both firms and individuals) will freely reveal their innovations and research (e.g. Harhoff et al., 2003) and that new forms of legal instrument have been developed that are designed to prevent rents being extracted from innovations (e.g. see Tang, 2005). Indeed, it has also been

---

market will also be a user. Given the right circumstances, individual users will innovate around their mass-produced consumer product in just the same way that a firm will innovate around a business-business process technology that they have acquired. As we know, the term 'user innovation' indicates the source from which the innovation has emerged: user innovation comes from the individual or firm that has acquired a product in order to use it, and not from the firm that has supplied it.

argued that the dominance of patents within innovation systems also presents a barrier for the detailed examination of alternative models of intellectual property development and value creation (e.g. see Gault and von Hippel, 2009).

The development of a broader definition of innovation that goes beyond the linear model is an important project that will require a tremendous amount of work at both the national international level. Within the UK the National Endowment for Science, Technology and the Arts (NESTA) has been a major actor in this respect and has published a series of studies (NESTA, 2006; 2007) that highlighted the inadequacies of traditional innovation indicators and explored forms of innovation that have remained 'hidden' and do not appear in official statistics. Internationally, the OECD is also in the process of developing a new innovation strategy that is due to be delivered, in its initial form, by 2010 and is likely to be a major shift in the international standards that govern the measurement and management of national innovation policy. Initiated in 2007, the new OECD innovation strategy is designed to provide an improved understanding of how innovation occurs in a globalised market for science and technology and assist policy makers to better measure and promote innovation. The strategy is being developed in the context of a broader definition of innovation, of which user innovation is a part, but also to reflect the complex innovation ecosystems that now exist:

> 'Developing such a strategy poses a significant challenge because of the complex nature of innovation...Knowledge is an important ingredient of innovation. It crosses the boundaries of institutions, both public and private, and comes from many sources, not just formal research and development (R&D) units. However, in a world of modularity and flexible platforms for technologies and practices, users are more able to improve and augment their systems, creating a web of knowledge that can change the behaviour of other users or of suppliers. This is a complex process, a system of innovation'

> (Gault and Huttner, 2008)

An important aspect of the OECD's strategy is the development of new indicators and metrics that will be required to provide a quantitative account of the broader definition of innovation and will represent a re-framing of policy away from the linear model. This is no small matter and will require the development of new generation of robust metrics and indicators that will adequately capture forms of activity, like user innovation, that have not been measured before. This is important, both because their absence is a limitation in changing models and frameworks (Godin, 2006), but also because policy action requires a strong evidence base. For example, a recent UK policy document (BERR, 2008) noted that proposals for Government intervention need to be evidence based, address a demonstrated market failure, show a positive impact assessment, and be proportionate, accountable, consistent, transparent and targeted. The collection of data to build appropriate metrics and indicators with which to provide the level of evidence required is meet these criteria is therefore an essential element of the policy understanding of the wider definition of innovation.

The Growing Policy Relevance of User Innovation Research

The need to engage with a wider definition of innovation has given existing user innovation research new relevance and created a demand for a new form of policy-focused research. User innovation research now forms part of the new, much broader body, of policy informed innovation research. This form of research is driven by different needs, asks different types of question and works to different timescales than many forms of academic research. Issues like market failure become important, and the need to provide substantial (preferably quantitative) evidence of cause, effect and impact places a different form of rigour on the research process. Drawing user innovation into policy is important, but sits alongside the many complex and interconnected challenges faced by government, as this extract from a recent UK policy document illustrates:

> 'Innovation is essential to the UK's future economic prosperity and quality of life ... Science and technology are a vital source of innovation. Innovation happens across the private, public and third sectors. Businesses are increasingly engaging in 'open innovation', reaching outside their walls for ideas. Users are innovating independently and in partnership with organisations, creating the demand for new products and services... This changing face of innovation is challenging businesses, Government and wider society to think differently if we are to have a successful economy and society over the next decades. Harnessing all the different types of innovation across all sectors is essential if we wish to create the conditions in which our economy can prosper.'

(DIUS, 2008)

User innovation is an important part of this emerging policy agenda, and is likely to grow in importance as new research on metrics and indicators develops. But this recent recognition of user innovation's policy relevance builds on large body of research that shows that user activity is a significant, widespread and valuable source of innovation in both industrial and consumer goods. This large body of work is reviewed by von Hippel (von Hippel, 2005), and elsewhere in this volume, and shows that significant proportions of users innovate in fields as diverse as software (Urban and von Hippel, 1988; Morrison *et al.*, 2002; Franke and von Hippel); medical instruments – 22% (Lettl *et al.*, 2006); sporting equipment (Luthje *et al.*, 2005). Research has also explored a series of features of user innovation including horizontal networks (von Hippel, 2007); how firms make use of Open Source (Dahlander and Magnusson, 2008) and hacker communities (Flowers, 2008), the role of toolkits in user innovation (e.g. von Hippel and Katz, 2002) and the role of users as co-developers (Jeppesen and Molin, 2003). The observation that that many users will apparently freely reveal their innovations has also been explored (von Hippel ad von Krogh, 2003), as have the voluntary information spillovers that emerge (Harhoff *et al.*, 2003), and the social welfare gains that result from this behaviour (Henkel and von Hippel, 2004). This large and vibrant body of research not only provides the foundation for the growing policy relevance of user innovation, but also a feedstock for a new wave of policy focused research.

This body of research provides extensive evidence of forms of innovation activity that lay outside the linear model view of innovation and raises many potential issues for policy including IP, fair use, and social welfare. The privileging of mechanisms to measure and support formal over informal systems of innovation appears incomplete and partial in this context and raises questions for other policies that support innovation in a more indirect manner. It is likely that these 'framework conditions' (national institutional and structural factors that surround innovation activities, including economic, legal, financial, and educational factors) will also need to be reappraised in the context of a broadening of the definition of innovation.

Policy Initiatives

This section will provide a brief overview of the main national initiatives that have focused on promoting or measuring user innovation.

Denmark — National programme for user-driven innovation

The Danish Government is arguably in the vanguard of national policy initiatives and in 2007 launched a targeted programme on user-driven innovation that will run for three years to 2010, with an annual budget of DKK100m (DEACA, 2007). The main intention of this programme is to strengthen methods for the diffusion of user-driven innovation in both the private and the public sectors. Initiatives within this programme include the creation of a fund to support new user-driven projects, and the establishment of research centres to further explore this phenomenon.

Although it draws heavily on the work of von Hippel, the Danish programme takes much wider definition of user involvement I innovation. For example, although one aspect of this wide-ranging and ambitious programme is designed to result in the development of new products, services and ideas, it is not solely focused on user innovation per se. Other aspects of the programme are intended to contribute to increased growth in participating firms, and to improvements in the efficiency and user satisfaction of participating public bodies. Another key part of the programme is the upgrading in the qualifications of employees in both the public and the private sector firms that participate in the initiative. The programme combines a series of cross-cutting strategic initiatives concerned with industrial, social and welfare issues with a series of regional activities that focus on the effective dissemination of experiences and methods. A final aspect of the programme was to commission special projects focused on user-driven innovation (DEACA, 2007).

Canada — measuring user innovation in manufacturing

Canada's national statistics agency, Statistics Canada, was one of the first bodies to systematically examine aspects of user innovation using large-scale surveys of industrial and consumer activity. The first large-scale survey providing evidence of firm-level user innovation was the 1998 Statistics Canada survey of Advanced Manufacturing Technologies. Designed to investigate the extent to which Canadian manufacturing firms use advanced technologies in their processes at the unit or 'plant' level, the survey was based on a sample of 4,200 firms with more than 10 employees. The study revealed that although the preferred method of acquisition of advanced manufacturing technology was by simply purchasing off-the-shelf equipment (84%) (Sabourin and Beckstead, 1999), a significant fraction (26%) of the firms reported that they had created their technologies by either customising or significantly modifying an existing technology. Further, a significant proportion (28%) had developed new technologies in-house (Arundel and Sonntag, 1999). This was remarkable finding as it indicated that a significant proportion of Canadian manufacturing firms were user innovators and that process technology modification and creation — key aspects of firm-level user innovation — were widespread.

This finding was explored further in the 2007 survey of Advanced Manufacturing Technology which included separate follow-up studies for: i) firms who modify existing technologies; and ii) firms that develop brand new technologies. These two surveys were designed to examine a range of issues including the way in which technologies were created or modified, how the innovations were diffused, and firm expenditures. These two surveys (reported in Gault and von Hippel, E., 2009) indicated that a significant fraction of firms (35% of modifying firms and 50% of firms that develop brand-new technologies) were engaging in user innovation on an ongoing basis. The survey also found that although almost all firms (98%) funded these innovation activities from their own resources, the mechanisms differed and formal R&D budgets were not necessarily the main source of funds. Interestingly, the most important source of funds for firms that modify technologies was the maintenance budget (52%), whilst for firms that develop new technologies the most important source was the R&D budget (54%).

Gault and von Hippel (ibid) also report that firms often take measures to protect their innovations (Modifiers: 46%; New Technology Developers: 60%), and that the majority don't share their modifications or newly developed technologies with other firms (Do not share — Modifiers: 83%; New Technology Developers: 81%). However, of the firms that do share their modified or newly developed technologies, a significant proportion do so at no charge (Shared at no cost — Modifiers: 75%; New Technology Developers: 47%), providing a voluntary spillover of valuable proprietary knowledge.

Taken together the two surveys indicate that proportion of Canadian manufacturing firms that adopt technologies by modifying or developing them is broadly the same with about 20% in each category (ibid). Put another way, around 40% of Canadian manufacturing firms are engaged in some form of user innovation.

The Netherlands — measuring user innovation in SMEs

In a neat complement to the work of Statistics Canada, which focused on enterprises with more than 10 employees, recent work in The Netherlands (De Jong and von Hippel, 2008) has examined user innovation in SMEs. This survey, organised by EIM Business and Policy Research on behalf of the Dutch Ministry of Economic Affairs, was based on 2,416 responses from firm with between 1–100 employees. The survey focused on user creation and user modification of '...existing techniques, equipment or software ...' excluding product modifications on behalf of customers and, in common with the Statistics Canada survey, also collected data on diffusion and expenditures. However, in order to explore the similarities between process and user innovation the survey also examined firm-level process innovations.

At a headline level the survey revealed that around 22% of Dutch SMEs were user innovators, (Modifiers: 18%; New Technology Developers: 4%). On a sector level it was found that manufacturing firms had the highest proportion of user innovators (Modifiers: 31%; New Technology Developers: 11%), with Business Services (Modifiers: 21%; New Technology Developers: 6%), and Farming (Modifiers: 20%; New Technology Developers: 4%) making the top three sectors. The lowest reported level of user innovation was by the Lodging and Meals sector (Modifiers: 10%; New Technology Developers: 1%). The study also found a strong positive correlation between firm size and the propensity to engage in user innovation, a finding that is in line with the observation on the earlier Statistics Canada study (Arundel and Sonntag, 1999) that there is a clear upward trend in the internal capabilities required by firms to undertake more complex forms of innovation.

In examining the relationship between user and process innovation de Jong and von Hippel (op cit) found that user innovation was only a partial subset of the level of process innovation. They found that a small proportion (10%) of the activities of the SMEs classified as user innovators do not qualify as process innovations and that user innovation appears to throw new light on apparently 'hidden' innovation (NESTA, 2007).

A further survey by EIM Business and Policy Research of high-technology SMEs in 2007 also explored the incidence and nature of user innovation. Based on responses from 498 high-technology SMEs drawn from a panel sample that was composed of firms with 1–100 employees operating in knowledge-intensive activities including, chemicals, rubber and plastics, machinery and equipment, IT and software developers, and commercial R&D services firms. In this more specialist sample it was found that 54% of the sample had been engaged in some form of user innovation in the previous three years. User modification was also lower than user creation (Modifiers: 32%; Creators: 41%), probably reflecting the high level of firm capability in the sample. In terms of the diffusion of the innovations created it was found that only 13% had been protected in some form, with 25% of all user innovations having been transferred to firms higher up the supply chain. Of those innovations that were transferred nearly half (48%) were given away at no cost, with the bulk of the remainder (39%) being transferred on the basis of some form of informal payment. Just 13% of the user innovations that were transferred were the subject of formal royalty agreements.

The UK — Measuring user innovation in firms and individual consumers

The UK has begun to engage with the policy implications of user innovation as part of a wider programme of work aimed at developing a better understanding of the broader definition of innovation. Within this larger programme of policy research activities user innovation has also been explored in the context of UK industries (e.g. Flowers *et al.*, 2008) and its implications for policy seriously examined. The publication of the Innovation Nation White Paper in 2008 (DIUS, 2008) was significant in this respect in that it was framed around a broader definition of innovation and recognised the importance of 'hidden innovation' to the UK's economy and society. The importance of users as a source of innovation was recognised and user innovation became a leitmotif throughout the White Paper that laid out the next phase of UK innovation policy.

One outcome of the White Paper was the establishment of a new UK innovation Research Centre specifically charged to examine wider forms of innovation, including user innovation. Another outcome was the commitment to create an Innovation Index that could include all forms of innovative activity, including those that have previously been overlooked in official statistics. The Innovation Nation White Paper recognised that a great deal of technical work needed to be done to develop a set of metrics and indicators to enable a more detailed policy understanding of wider forms of innovation. NESTA was charged with developing an Innovation Index to begin the process of measuring non-traditional forms of innovation, including user innovation. Within this programme of work one project is designed to measure user innovation in the UK and design a set of indicators that may form part of an Innovation Index that covers the broader definition of innovation. The initial version of the Innovation Index was published in late 2009, with a more comprehensive system planned to be in place by 2010.

The UK work to measure user innovation drew heavily on the earlier Canadian and Dutch studies in its firm-level survey, but also explored the extent of user innovation at the level of the individual consumer. The research employed a telephone survey to examine user innovation within a structured sample of 1004 firms with between 10–250 employees across 15 sectors. The study found that 15% of the firms have engaged in some form of user innovation over the previous three years (Modification: 10.3%; Creators: 8.6%). On a sector level there was a wide variation, with Software and IT services being the most active user innovators (50%) with Other Creative Activities including design and architecture (24.7%) Other Manufacturing (23.2%) and Aerospace and Automotive (20%) reporting high levels of activity. As in previous surveys larger firms were found to be the more active user innovators, probably reflecting the broader and deeper scale and scope of their resources. It was also found that the level of user innovation in software was around twice the level that was found in hardware.

The UK consumer-level survey was the first attempt to map the size and shape of user innovation within the general consumer population and was a move into new ground in user innovation research. This study took a two-stage approach towards measuring user innovation in a general consumer population. The first stage was designed to inform the design the second more in-depth survey and examined several aspects typically associated with user innovation including content sharing and product modding. The responses from the first omnibus survey of 2109 individuals aged over 15 showed that 34% share content online, with that number increasing to over 50% amongst the under 35s. A much smaller percentage — 10% — reported modifying software by changing the computer code or using a programming language. However, 20% reported modifying things like the tools, toys, sporting equipment, cars and household equipment they use in everyday life, with 13% of respondents saying that they have created their own products from scratch. These responses were used to inform the design of the sample for the second survey.

The second study, an in-depth telephone survey of 300 consumers was designed to enable responses to be verified and overall estimates for user innovation amongst UK consumers to be calculated. The results showed that 8% (Modifiers: 5.2%; Creators, 4.4%) of the consumer population in the UK had engaged in some form

Table 1. Summary of results from user innovation measurement studies.

Study	Modifier %	Creation %	Overall %	Sample
Statistics Canada 1998[3]	26	28	—	4,200 manufacturing firms, >10 employees.
Statistics Canada 2007[4]	21	22	43	1,219 manufacturing firms, >10 employees
Netherlands SMEs, 2007[5]	18	4	22	2,416 SMEs, 1–100 employees
Netherlands Hi-tech firms, 2007[6]	32	41	54	498 Hi-tech SMEs, 1–100 employees
EU Firms, 2007[7]	30	—	—	5238 firms across 27 EU member states, 20+ employees
UK Firms, 2009[8]	10	9	15	1004 firms across 156 sectors, 10–250 employees
UK Consumers, 2009[9]	5	4	8	Structured sample of 2109 consumers aged 15+

of user innovation in the previous three years. Applying this result to the population as a whole indicated that something like 3.8 m UK consumers had been engaged in some form of user innovation in the previous three years.

The EU — towards measuring user innovation across 27 countries

The 2007 Innobarometer survey was the first attempt to examine aspects of user innovation activity within the European Community. The annual Innobarometer surveys are based on a random sample of companies employing >20 persons identified from Dunn & Bradstreet data for each country. The sample is stratified according to size and sector and the 2007 survey obtained responses from 5238 firms. The survey revealed that 30% of innovative firms innovated by modifying their process technologies, with larger firms being more active in this form of user innovation

---

[3] Arundel and Sonntag, 1998
[4] Schaan and Uhrback, 2009
[5] de Jong and von Hippel, 2008a
[6] de Jong and von Hippel, 2008b
[7] Flowers, Sinozic, Patel, 2009
[8] Flowers, von Hippel, de Jong, Sinozic, 2009
[9] ibid

activity. Interestingly, firms engaged in forms of user innovation emerged from this study as 'super-innovators' as, compared to other innovative firms, they were more likely to introduce new products, processes or services. They are also more likely to initiate new organizational methods and a higher proportion of user innovators carried out both intra and extra mural R&D and also apply for patents. This study suggests that user innovation activity may, in itself, be an indicator of firms that are highly innovative.

This collection of studies tend to indicate that forms of user innovation are widespread and may be a highly significant source of innovative activity in modern developed economies. Making direct comparisons of the results of the various surveys that have taken place is difficult due to the differing samples and methodologies that have been applied. However, these surveys provide strong evidence that user innovation is a measurable phenomenon that occurs across many industrial sectors, but is more prevalent amongst larger, manufacturing and hi-technology firms.

Towards Developing User Innovation Indicators

The measurement initiatives in Canada, The Netherlands and the UK are all aimed at providing the basis for informed policymaking: the development of reliable and robust mechanisms for measuring user innovation at the level of the firm.

The metrics collected on the scale of firm level user innovation, together with the costs, value, transfer of IP, and spillover effects can all be drawn upon by policymakers and collected together to become indicators of this activity. Such indicators may be thought of as quantitative representations designed to provide summary information concerning a particular topic, and are widely used in policy. They tend to be collections of individual statistics that show how an aspect of performance or behaviour changes over time (Godin, 2006, p107). In general, indicators are developed to be used and the policy community is the target (Gault, 2007). Typically, they are based on a model that '... explicitly tests some assumption, hypothesis or theory' (Holton, 1978). In short, indicators are designed to provide information about an important aspect of an economy or society, and they reflect a guiding hypothesis that frames their creation, interpretation and use.

The OECD has been the pre-eminent source of Science, Technology and Innovation (STI) indicators for over 50 years which '...measure and reflect the science and technology endeavour of a country, demonstrate its strengths and weaknesses and follow[s] its changing character notably with the aim of providing early warning of events and trends which might impair its capability to meet the country's needs' (Sirilli, 2006). As we have seen the current set of metrics and indicators is based on a linear model of innovation that privileges role of science and technology within the innovation system (e.g. NESTA, 2007; OECD, 2007). As a result certain forms of activity, including user innovation, have been outside the 'official' definition and have been excluded from mainstream statistics and indicators. The move towards a wider definition of innovation that includes user innovation (as well as many other forms of innovation hitherto overlooked in official statistics) means that a new set of indicators will need to be developed to supplement the large number of indicators already in use.

In the context of user innovation, this development is still at an early stage but research underway in Canada, The Netherlands and the UK may point to their eventual shape and composition. It is beginning to be clear that it is possible to reliably collect representative data on firm-level user innovation on a range of issues, including:

- The incidence of user innovation, including hardware and software modification and creation, together with specific examples;
- The part played by supplier and other user firms in the user innovation activity;
- The expenditures in terms of time and money associated with a particular user innovation;
- Whether the innovation is protected or freely revealed;
- How it was shared, if at all;
- Its adoption by suppliers or other users;
- The level of public-sector R&D support that the user innovation benefitted from, if any.

Collecting data on the scale and scope of firm-level user innovation will provide a major insight into its scale and scope and provide a basis for sectoral analysis and examination of size effects. It will also enable the impact of technological capability on user innovation to be assessed. Linking firm-level user innovation to data on reported R&D will enable policy to better understand the extent to which current R&D indicators reflect the extent of wider innovation that takes place and enable the development of composite innovation indicators that take account of all forms of innovation activity.

Developing indicators that provide insights into the ways in which firm-level users transfer innovations to suppliers or to other users will provide a better understanding on the transfer mechanisms beyond patents and licensing, of which a great deal is known. It will also enable sector and size effects to be better understood and the effect of a firm's location in the value chain to be assessed. Taken together these will enable policymakers to develop a more informed appreciation of a wider range of mechanisms for innovation transfer and to develop more nuanced policy positions concerning IP. They will also enable policy to better assess if R&D support measures are correctly targeted and provide a basis for policy measures to encourage or support innovation in leading or lagging firms/sectors.

Developing indicators that can capture the nature of innovations by individual end consumers, the scale and scope of innovative activity and the transfer and diffusion of their innovations will enable policymakers to more fully appreciate the part played by individuals, and communities of individuals, in innovation. Individual consumers are currently excluded from the 'official' innovation that is reflected in R&D statistics, innovation surveys and other indicators. Yet it is clear that individuals, and communities, can be highly innovative, can have a significant impact on sectors (e.g. software, sporting goods) and give rise to wholly new forms of organisation and collective output (e.g. open source software).

Indicators like those outlined above will provide policy with a better understanding of the broader innovation ecosystem and provide a basis for the development of a measure of wider R&D that includes user innovation and other forms of 'hidden' innovation. They will provide the basis for a more detailed understanding on the positive spillover effects, both economic and societal, that accrue from firm-level, individual and community user innovation activity. This will enable a more detailed exploration of the boundaries of user innovation and enable the development of nuanced and targeted policy in this area.

The broadening of the definition of innovation also raises questions about the guiding assumptions concerning IP, fair use and the parts users can productively play in innovation. For example, there is a great deal of evidence across many literatures that users will often seek to 'mod' existing products. This has been observed in many sectors and although some firms and sectors seek to embrace this, but others lobby to limit this activity and criminalise it. The shift away from a linear model of innovation that, in policy terms, positions users simply as consumers to a broader definition of innovation that recognises them as innovators raises questions for this stance. The move towards a broader definition of innovation, and the creation of a set of indicators and metrics that measures the scale and scope of this activity, will provide the evidential base required to begin to understand the social and economic value that is created and provide the basis for a more informed debate on issues like 'fair use'.

The Developing User Innovation Policy Research Agenda

The move towards the adoption of a wider definition of innovation opens up new areas for policy-focused user innovation research and enables the reassessment of existing research findings. The systematic collection of firm and consumer-level data concerning user innovation activity will enable the more detailed understanding of user innovation to be developed at the firm, sectoral and economy level. This will extend the work on individual user communities and provide an opportunity to better understand the dynamics of user activity at a macro level. It will enable both the 'vertical' user innovation ecosystem that exists around products, firms, and sectors to explored and the importance of spillovers that come from 'horizontal' user innovation ecosystems to be better understood. New theoretical approaches will be very useful in this respect and draw user innovation into the mainstream of economic analysis.

It will also be important to better understand the contingent factors surrounding user innovation, the barriers that must be overcome and the factors that facilitate it. This will enable policymakers to begin to frame the specific policy responses and instruments that will enable policymakers to maximise the economic and social benefits that accrue from user innovation. To this end, research will need to better understand the dynamics of specific forms of user innovation activity and its link to entrepreneurial activity and the formation of firms. There is interesting work on the barriers to user innovation (e.g. Braun and Herstatt, 2009) but more quantitative work needs to be done to better inform how policy should react to this phenomenon. It will also be important to understand the conditions under which no-cost transfers take place and the extent and limits of free revealing.

The development of user innovation policy research will draw heavily on the existing body of user innovation research for its theoretical framing but is likely to be based on large scale surveys of consumer and firm behaviour. Case studies will remain highly important in exploring new aspects of user innovation and for developing new theoretical insights and refining existing theories, but quantitative approaches are likely to predominate in this new domain. Similarly, econometric models of user activity will need to be developed to explore spillover effects and assess the impact of policy actions on the outcomes of user innovation. As a result, the measurement and modelling of user innovation populations is likely to emerge as a new stream of research that could be highly influential in the policy context.

References

Arundel, A and V Sonntag (1998). Patterns of Advanced Manufacturing Technology (AMT) Use in Canadian Manufacturing: 1998 AMT Survey Results, Research Paper 88F0017MIE (Ext. r. no.). Canada: Statistics Canada.

Arundel, A, C Bordoy and M Kanerva (2008). Neglected innovators: How do innovative firms that do not perform R&D innovate? INNO-Metrics Thematic Paper.

BERR (2008). Consultation on legislative options to address illicit peer-to-peer (P2P) file-sharing, July.

Braun, V and C Herstatt (2009). User-Innovation: Barriers to Democratization and IP Licensing, Routledge.

Dahlander, L and M Magnusson (2008). How do firms make use of Open Source Communities? *Long Range Planning*, 41(6), 629–649 (December 2008).

DEACA (2007). Programme for user-driven innovation, Danish Enterprise and Construction Authority, www.deaca.dk/userdriveninnovation, accessed May 1st 2009.

de Jong, J and E von Hippel (2008). User innovation in SMEs: incidence and transfer to producers. *SCALES working paper*, H200814.

DIUS (2008). Innovation Nation, HMSO, Cm 7345.

Flowers, S (2008). Harnessing the hackers: The emergence and exploitation of Outlaw Innovation. *Research Policy*, 37, 177–193.

Flowers, S, A Grantham, J Mateos-Garcia, P Nightingale, J Sapsed, P Tang and G Voss (2008). *The New Inventors; How Users are Changing the Rules on Innovation*, National Endowment for Science Technology and the Arts (NESTA), July.

Flowers, S, T Sinozic and P Patel (2009). Prevalence of User Innovation in the EU: Analysis based on the Innobarometer surveys of 2007 and 2009, INNO-Metrics Thematic Paper.

Flowers, S, von E Hippel, J de Jong and T Sinozic (2009) Measuring User Innovation in the UK, NESTA.

Franke, N, E von Hippel and M Schreier (2006). Finding commercially attractive user innovations: a test of lead-user theory. *Journal of Product Innovation Management*, (23), 301–315.

Gault, F (2007). Science, Technology and Innovation Indicators: The Context of Change, in Science, Technology and Innovation Indicators: Responding to Policy Needs, OECD.

Gault, F and S Huttner (2008). A cat's cradle for policy, *Nature*, 455/25, 462–463 September.

Gault, F and E von Hippel (2009). The prevalence of user innovation and free innovation transfers: Implications for statistical indicators and innovation policy, MIT Sloan School of Management Working Paper 4722–09.

Godin, B (2006). *Measurement and Statistics on Science and Technology: 1920 to the present*. Routledge.

Harhoff, D, J Henkel and E von Hippel (2003). Profiting from voluntary information spillover show users benefit by freely revealing their innovations. *Research Policy*, 32, 1753–1769.

Henkel, J and J von Hippel (2004). Welfare implications of User Innovation. *The Journal of Technology Transfer*, 30, 73–87.

Holton, G (1978). Can Science be Measured? In *Towards a Metric of Science: The Advent of Science Indicators*, Elkana et al. (eds.), John Wiley & Sons.

Iansiti, M and R Levien (2004). *The Keystone Advantage: What the New Dynamics of Business Ecosystems Mean for Strategy, Innovation, and Sustainability*. Boston, Mass: Harvard Business School Press.

Jeppesen, LB and MJ Molin (2003). Consumers as co-developers: learning and innovation outside the firm. *Technology Analysis & Strategic Management*, 15(3), 363–383.

Klein, SJ and N Rosenberg (1986) An Overview of Innovation, In Landau, R,. Rosenberg, N. The Positive Sum Strategy: harnessing technology for economic growth, National Academy of Engineering.

Lettl, C, C Herstatt and HG Gemuenden (2006). Users' contributions to radical innovation: evidence from four cases in the field of medical equipment technology. *R&D Management* 36, 3.

Lundvall B and S Borras (2005). *Science, Technology and Innovation Policy*, in The Oxford Handbook of Innovation, OUP.

Luthje, C, C Herstatt and E von Hippel (2005). User-innovators and "local" information: The case of mountain biking. *Research Policy*, (34)6, 951–965.

Morrison, PD, JH Roberts and DF Midgley (2004). The nature of lead users and measurement of leading edge status. *Research Policy*, 33(2), 351–362.

NESTA (2006). The Innovation Gap: Why Policy needs to reflect the reality of innovation in the UK. Research Report, October.

NESTA (2007). Hidden Innovation: How innovation happens in six 'low innovation' sectors. Research Report, June.

OECD (1994). The Measurement of Scientific and Technological Activities: Using Patent Data as Science and Technology Indicators — Patent Manual.

OECD (1996). The Measurement of Scientific and Technological Activities: Proposed Guidelines for Collecting and Interpreting Technological Innovation Data — Oslo Manual.

OECD (2002). The Measurement of Scientific and Technological Activities: Proposed Standard Practice for Surveys on Research and Experimental Development — Frascati Manual.

Sabourin, D and D Beckstead (1999). *Technology Adoption in Canadian Manufacturing, Survey of Advanced Technology in Canadian Manufacturing*, Statisitics Canada, August.

Sainsbury (2007). *The Race to the Top: A Review of Government's Science and Innovation Policies*. HM Treasury, HMSO.

Schaan, S and M Urbach (2009). Measuring user innovation in Canadian manufacturing 2007, statistics Canada. *Science*, 48(7), 821–833, July.

Sirilli, G (2006). ISSiRFA, Italiy, Developing science and technology indicators at the OECD: the NESTI network, 16 p. Presented at the ENID / PRIME International Conference "Indicators on Science, Technology and Innovation: History and New Perspectives", Lugano, Switzerland, 16–17 November 2006.

Tang, P (2005). Digital copyright and the "new" controversy: Is the law moulding technology and innovation? *Research Policy*, 34, 852–871.

Urban, GL and E von Hippel (1988). Lead user analyses for the development of new industrial products. *Management Science*, 34(5), 569–582 (May, 1988).

von Hippel, E and R Katz (2002). Shifting Innovation to Users *via* Toolkits, Management

von Hippel, E and G von Krogh (2003). Open source software and the "private-collective" innovation model: Issues for organization science. *Organization Science* 14(2), 208–223.

von Hippel, E (2005). *Democratizing Innovation*. Cambridge, MA: MIT Press.

von Hippel, E (2007). Horizontal innovation networks — by and for users. *Industrial and Corporate Change*, 16(2), 293–315.

THE FREEDOM-FIGHTERS: HOW INCUMBENT CORPORATIONS ARE ATTEMPTING TO CONTROL USER-INNOVATION

VIKTOR BRAUN
Massachusetts Institute of Technology, USA

CORNELIUS HERSTATT
Hamburg University of Technology, Germany

Introduction

The classical manufacturer-driven view of the technological progress that was prevalent both in academic minds and in industrial practice for a large part of the 20th century, seems to have become anachronistic. Technological innovation and the flow of capital have become too complex and too fast for the previously successful model of accumulating huge R&D facilities, occupying these with the most talented employees available and thereby developing globally successful products in-house. It has become obvious that in order to remain innovative in today's dynamic and technocratic global market, corporations need to collaborate with external parties such as corporations, universities, governmental agencies, suppliers and even users (Biemans, 1991; Chesbrough, 2003).

There is a growing evidence that companies are increasingly using outside resources in their innovative efforts, as suggested by the open-innovation paradigm (Chesbrough and Kardon Crowther, 2006; IRI, 2007). Open-innovation refers to the principle that companies should leverage external discoveries and search for companies who have the best business model for a specific technology (Chesbrough, 2003). In recent years, we have indeed witnessed an increasing tendency of companies not only to outsource their production function but also parts of their R&D activities (Howells, 1999; Engardio and Einhorn, 2005). Surprisingly, this does not imply that the open-innovation paradigm refers to a model in which everybody is free to participate in the innovation process. On the contrary, many of the most successful companies of our times, such as the Microsoft, ExxonMobil or General Electric, are certainly not known for the openness of their innovation activities.

Users have been especially affected by the reluctance of many companies to integrate them into their new product development processes. This is despite an increasingly prominent stream of research — which we believe, should be incorporated into the open-innovation literature — that has demonstrated that users have been a significant source of innovation over the last few centuries. Farmers, historically probably the largest group of user-innovators, have improved agricultural yields for thousands of years (Douthwaite, 2002). Athletes have created an abundance of different sporting activities such as, most recently, mountain biking (Lüthje et al., 2005), kitesurfing (Tietz et al., 2005) and rodeo kayaking (Hienerth, 2006). Medical doctors have developed numerous novel medical devices and treatments (Biemans, 1991; Lüthje, 2003; Lettl et al., 2006). Innovative users have also been encountered in areas as distinct as transport (Lüthje, 1999), oil refining (Enos, 1962), electronic circuit boards (Nagel, 1993), libraries (Morrison et al., 2004) or construction equipment (Herstatt and von Hippel, 1992).

Computer users, who by tinkering with open-source software are re-programming the digital world, have been the most visible user-innovators in recent years (Berners-Lee, 2000; Shapiro, 2000). Successful new programs such as Apache, Sendmail, Bind, Firefox or Linux illustrate the enormous potential of user mass-collaboration (Feller and Fitzgerald, 2002). The explosion in information- and communication-technologies has enabled the instantaneous quasi-costless sharing of software and lead to global virtual user (innovation) communities (Tietz et al., 2006). By dramatically facilitating the independent exploitation and distribution of users' developments, the software development process has been re-shaped.

This potential for users to increasingly tailor-make products to suit their unmet needs has been highlighted in this literature. Prahalad and Ramaswamy (2004), for instance, argued that a silent transformation is afoot leading to the co-creation of value between the consumer and corporations. Neff and Stark (2004, p. 175) describe their vision of a "permanently beta" business environment, where technologies are constantly in a state of flux and subject to "a process of negotiation [with] users". Zittrain (2005, p. 19) implicitly refers to the potential exponential growth of user-innovators in this newly created, "Galapagos-like ecosystem of software designers". Ramirez (1999, p. 59) even suggested that "systems must be installed to track what happens to the customers' value creating". Tapscott and Williams (2006) in their bestselling work, *Wikinomics*, proclaim the beginning of a new era in which user mass-collaboration will revolutionise our economic processes. Professor von Hippel, who started the user-driven innovation studies in the late 1970s, brings these views to their logical conclusion and argues that innovation is being "democratised";

"Users of both products and services — both firms and individual consumers — are increasingly able to innovate for themselves" (von Hippel, 2005, p. 1).

Despite the enormous potential of users, the literature has at times, potentially slightly exaggerated the thrust for openness and underestimated the resistance to the change of incumbent corporations. There are still many corporations that do not actively involve users as innovative agents. They merely collect the market feedback, i.e., need-related information, and ignore the possibility of users offering valuable solution-related information or even developing product improvements themselves. There are more reasons for this behaviour than corporate resistance to change. Strategic and financial considerations can be so strong that, at least in the short-run, it can be sensible for some incumbents to actively limit what Edward Felten terms, "the freedom to tinker" (www.freedom-to-tinker.com).

Exploring this frequently overlooked phenomenon is of utmost importance due to the welfare-enhancing aspects of user innovations: (i) they tend to complement manufacturer innovations, filling small niches of previously unmet high needs; (ii) by decreasing the information asymmetries between manufacturers and users, they can potentially lead to more balanced economic exchanges among the parties and (iii) and as they are much more frequently and freely revealed than manufacturer innovations, they can contribute to the establishment of an information common (Henkel and von Hippel, 2005). User innovations, furthermore, allow high R&D costs to be partitioned among numerous parties, who individually assume a small part of the temporal and material costs. Finally, unlike manufacturer innovations, of which according to von Hippel (2005), 70%–80% fail in the market place, user innovations actually solve usage-related problems.

A democratisation of user innovation would have the further benefit of socialising the risks of technological progress. Although technology has increased the quality and comfort of our lives in countless ways from global communication to feeding the world population, the societal risks of a further rise of the division between the affluent and the destitute should not be ignored (Mossberger *et al.*, 2003). The substitution of human labour by machines and robots, not only in the production of products but also increasingly in the field of services and one day, even in the creative realms such as R&D (Kurzweil, 1999; Joy, 2000; Pearson, 2005) constitutes a further noteworthy risk. If a growing number of users were in charge of the technologies and their improvement, these risks would seem to be acceptable. If, however, technological progress is in the hands of a few players, the potential of technocratic abuse could threaten the fabric of our society. As a brief example, consider the large-scale crackdown on freedom of speech in China, as recently demonstrated by their substantial limitation of online videos (Greenberg, 2008). The proliferation of user-innovations thus promises to lead to a socially more efficient innovation process.

We commence this paper with an explorative discussion of corporate incentives to restrain user-innovation. Subsequently, we scrutinise numerous legal, technological, economic and social barriers, some of which manufacturers can use to actively change users' cost and benefit expectations with respect to innovations. We demonstrate these empirically via a study in the realm of a computerised dentistry. In the section on "Further Freedom Fighters", we briefly attempt to generalise our findings by showing how companies from other fields burden tinkering users. We end with an explorative discussion of the conditions under which an exclusion of user-innovation may be sustainable.

Corporate Incentives to Prevent User-Innovation

As with any societal or economic change, there will be winners and losers. If a party is confronted with a potentially adverse change, resistance is likely even if such a change is inevitable and adaptation to the new circumstances would be more sensible. As the proposed democratisation would constitute a substantial re-distribution of economic resources, it is understandable that incumbent manufacturers may not be euphorically embracing such a paradigm change.

More precisely, the democratisation involves users usurping a key core competence of many manufacturers — the ability to innovate. Innovation — just look at the role the word plays in the annual reports of most companies — is a key source of corporate competitive advantage. "Growing the business through innovation", was rated as the number one problem since 2002 in the US' Industrial Research Institute's annual survey (IRI, 2007). While collaborating with users may, in many situations, be a promising strategy, it also has its risks and limitations (Holt, 1988; Dahlander, 2006). It is, therefore, crucial to note that it is a strategic decision whether or not it makes sense for a corporation to collaborate with users. At 3M, for instance, the lead-user method was first successfully employed, then dropped only to subsequently be re-employed again (Lilien *et al.*, 2002; Thomke, 2002). Different CEOs had different strategic foci. A Scandinavian software firm also stopped using the lead-user method despite its successful applications due to both organisational inertia and the large amount of resources required to entrench the method in its organisational routines (Olson and Baake, 2001). As Gassmann (2006, p. 223) succinctly summarises: "open innovation is not an imperative for every company and every innovator". He suggests a contingency approach in which companies strategically evaluate which problems are best solved via an open- and which via a more closed-innovation approach.

What are, therefore, the main reasons to erect hurdles to user-innovation? The essence of corporate behaviour is the quest to earn profits. This economic motivation is at the forefront of any reluctance to embrace user-innovation. Basic economic theory promulgates that by charging different customers different prices, corporations can maximise their profits (Klein, 1996). After market strategies of underpricing hardware, e.g., razors and over-pricing, after market components, such as razor blades, exist in a wide variety of markets (Peritz, 2002). In order to discriminate among the different types of users, incumbents have to exclude non-authorised re-use of their product. Microsoft, for example, attempts to limit the use of its software to one computer (Spanbauer, 2003). Mobile phone operators are keen to ensure that the telephones they sell are not used with another provider. In order to implement a market segmentation strategy and control their rent streams companies, thus, strive to "build electronic locks and fences around their properties" (Granstrand, 1999, p. 399).

The best example of a business model, which requires the control of aftermarkets is the printer industry. Printing companies often sell relatively cheap printers, intending to make their money by selling expensive ink cartridges (Wagstaff, 2002; Spring, 2003). Hewlett-Packard and its competitors have increasingly employed a market segmentation strategy in this US$26 billion industry (Heuer, 2006), to charge these more, who print more. As users, however, were unhappy with such high costs, they started to tinker with re-filling cartridges (Varian, 2002), which led to the rise of the cartridge re-filling industry (Lyra Research, 2005). As this form of user-innovation clearly threatens their business model, the printing giants, as we will see below, reacted aggressively.

The second principal reason for combating user-innovation is the strategic one of protecting a companies' intellectual property. In order to prevent their competitors from appropriating their know-how companies, therefore, they often try to increase the difficulty of reverse engineering. Car manufacturers, for example, often make car parts interdependent and in one closed, sealed unit so that when one part is broken, the whole unit has to be replaced. This makes repair exceptionally difficult for non-authorised parties (Pickler, 2002). Douthwaite (2002) generalises that corporations are more likely to introduce technologically complex products that are difficult to modify rather than simple solutions that can easily be shaped by users themselves.

Developing technologies, so that they cannot be modified or copied, is a strategy that is becoming increasingly important to combat piracy in developing countries. According to a research by the International Data Corporation (IDC), globally 35% of all the software was pirated in 2006, leading to alleged losses in sales of $40 billion (BSA and IDC, 2006). In China, apparently 82% of all the software was pirated (ibid). Regardless of whether these industry figures are exaggerated or not, they demonstrate the desire of the software industry to fight piracy. Lie (2003, p. 1) argues that this urge to prevent software piracy and enabling "distributed services such as banking transactions, on-line gaming, electronic voting, and digital content distribution ... without the risk of ... intellectual property theft," are crucial motivations for creating tamper-resistant products.

Further important reasons as to why companies intend to create non-modifiable products is the desire to evade legal liability, maintain quality control over their products (Rogers, 2003) and the possible adverse effects on reputation for damages that may result due to the products being modified in an imperfect manner or used for non-authorised purposes. User-developed products may, for example, not be in congruence with the companies' intended image. BMW, for example, found hackers, who had turned some of its cars into dangerous hotrod racing vehicles. Also, consider the case of a flatbed grain dryer, which was subsequently modified and re-constructed by the Vietnamese farmers. As these farmers made mistakes in reducing the airflows of the blower, the excessive hot dryers caused the grains to crack, considerably lowering its sale value (Douthwaite, 2002). The authors also made use of the experience that companies may be reluctant to allow users to modify their products or ask these about the product's quality due to the fear of sending an adverse message to the market.

In the realm of products employing software, and this refers to a continuously growing amount of goods (Graaf et al., 2003), another goal is to prevent viruses or malicious hackers from modifying the programs. This can result in a great loss to legitimate users and third parties (Berinato, 2003). An extreme example is provided by the Islamic terrorists who recently attempted to blow up various aeroplanes using their iPods to ignite home-made bombs (Focus, 2006).

Structural factors as well as the egocentricity of some manufacturers and diffusion agencies, which are frequently of the opinion, "that they know the best as to the form of the innovation that users should adopt" (Rogers, 2003, p. 184), also play an important role. Interacting too closely with users can furthermore "limit the firm's room for strategic action" (Dahlander, 2006, p. 41). Once a firm allows users to freely tinker with its products, a strategy reversal is likely to lead to very strong adverse reactions (Dahlander and Magnussen, 2005). The not-invented here syndrome may also be a reason why firms often struggle to embrace user-innovations. Douthwaite (2002, p. 57) mentions a further fascinating reason why corporations may be keen to impose certain hurdles to innovation, "erecting a barrier or two favours the self-selection of the right sort of people". In other words, raising the threshold of the users' ability to innovate, could allow corporations to identify lead-users with whom collaboration may be worthwhile.

To summarise, whether to allow, stimulate, burden or prohibit user-innovation in a specific realm will require a strategic assessment by the company in question. Riggs and von Hippel's (1994, p. 460) finding that in the scientific instruments industry, there is a "very significant tendency for users to develop innovations that have high scientific importance, and for manufacturers to develop innovations that have high commercial importance", seems to offer support for this view. In areas, where the commercial benefits are expected to be low, it would seem to make sense for manufacturers to stimulate users in their innovative activities. In areas, where the expectations of commercial benefit are very high, manufacturers will be much more inclined to control user-innovation. Whether user-innovation will be embraced or limited will also change over time. Take the example of Ford; in the beginning of the 20th century, it not only permitted users to tinker with their model T automobiles, but substantially benefited from it and used the slogan, "The Universal Car", quite literally to its advantage: model T automobiles were being used by tinkerers, mainly farmers, to run washing machines or to plough fields. Such an embracing of user-innovation seized when Ford started to sell not only model T vehicles but also trucks and tractors. As user-tinkerers were no longer needed to increase its popularity and instead threatened the sales of its new models, Ford warned that tinkering would damage the car's engine, drafted strong warranty provisions and exerted pressure on dealers not to sell truck or tractor kits to farmers (Kline and Pinch, 1994).

Corporate Weapons to Prevent User-Innovation: The Medical Device Industry

In a recently published article, the authors described the existence of the so-called barriers to user-innovation using the example of the seed industry (Braun and Herstatt, 2007). A framework based on Lessig's (1999) work was presented in which four interacting factors were responsible for shaping the overall innovation landscape and these are listed below:

- law
- market
- technology
- social norms

While some barriers, such as the overall inherent technological complexity, are out of their control, manufacturers can either directly erect barriers (such as through market power, contracts, Intellectual Property (IP) or technology), or like the legal and social environment indirectly shape these, by lobbying, for example. We will begin this section with a brief description of our object of study, the medical device industry. Subsequently, we will introduce the relevant barriers and obtain some empirical evidence of their existence in the German computerised dentistry industry. We will thereby describe how the manufacturer in question purposefully attempted to prevent user-innovation in this field.

The medical device industry is very much technology- and innovation-driven (Lotz, 1992). German medical device corporations, for example, earn over half of their revenues from products that are less than two-years-old (BMBF, 2005). Innovation in this realm is strongly inter-disciplinary with the areas of microsystems-technology, mechatronics, information- and communications-technology, such as computers and software, electronics, optical technologies, material sciences, nanotechnology and cell- and bio-technologies, all playing important roles (BMBF, 2005; Lettl et al. 2006).

We chose this field as a suitable object of study as it is characterised by a high degree of user-innovation. Lüthje (2003) found that 39.2% of the responding German surgeons previously had an idea of a new surgical device. While 13.7% stopped after having created a drawing of their idea, 22% either developed a prototype or a marketable product. In Shaw's (1985) study of 34 medical equipment innovations commercialised by the British firms, 53% were found to have been user-dominated. Despite the important role of users in this field, manufacturers also play an essential role. Biemans (1991), for example, found that over half of the product development processes in the Dutch medical device industry were initiated by a user or a third party, although as soon as it came to the concept of development stage, manufacturers started to dominate the process. He also found that across the whole innovation cycle, users were involved in around 50% of the activities, while manufacturers participated in nearly 90% of these activities. Lotz (1992, p. 9) concurs and states that "the predominant part of the innovative activity is located in manufacturing firms".

The global leader in computerised dentistry equipment is a medium-sized company that has a dominant market share in the niche market that it had created. Computerised dentistry refers to a computer-aided design/manufacturing (CAD/CAM) system that enables a dentist to design, manufacture and fit tailor-made ceramic restorations without any additional laboratory support (Pallesen and van Dijken, 2000). The specific system in question consists of two separate units, one for imaging and the other for milling. A computer-driven laser-lighted video camera captures three-dimensional digital images. A software subsequently determines the desired shape of the requisite inlays, onlays, crowns or partial crowns, which an attached robotic milling device then carves out of the ceramic blocks (Guinnessy, 1997). Currently, there are over 20,000 systems in clinical use around the world (Reiss, 2006).

We chose what we will refer to as CD due to the fact that we encountered a clash between the view proposed in the academic literature and the industrial perspective. Although there have not been any other user-innovation studies involving dentists, the existing studies (Biemans, 1991; Lotz, 1992; Lüthje, 2003; Lettl et al., 2006) together with the fact that dentists are skilled craftsmen indicate that dentists would be active innovators. The industrial view, however, as given by representatives of the manufacturer was that user-modifications were not common. The CEO commented that as we are dealing with "end-users" and not industrial intermediary-users, user-amendments of the CD were "unlikely". The product is simply "too complex", especially as these users "are not trained engineers ... [or] software programmers".

While originally we wanted to determine which view was more accurate in this area, this study enabled us to see how a manufacturer decided to erect high barriers for tinkering users. We specifically chose to investigate the German market for CD, as the technology was first commercialised by a German company, which for the last 20 years has further improved the technology, and as Germany is one of the most developed CD markets around the world (Wiedhahn, 2006).

The methodology that we employed was the mailing of a questionnaire to the German CD users and to follow this up with some interviews. In order to obtain non-biased and independent results, we approached the German Society for Computerized Dentistry (DGCZ) to send a questionnaire to their 1687 members. Given approximately 5000 CD users in Germany, this amounts to around 34% of the German users having been contacted. After conducting a pre-test, we mailed out the questionnaires in early December 2006. We obtained 283 responses to the mailing, which constituted an overall response rate of 16.8%.

The primary aim of the questionnaire was to locate user-modifications (including third-party modifications), which are either changes made to the hardware or software or new uses of the technology made *independent* of the manufacturer (Rice and Rogers, 1980). Any new software update or complementary gadget made or provided by the manufacturer was excluded. For everyone who responded affirmatively, we consulted some experts, the manufacturer, as well as, if possible, the respondent via telephone in order to determine whether or not the mentioned act constituted an actual modification.

Determining the need to innovate is a pre-requisite to any study of this kind. If the user is fully satisfied with the product in question, it is obvious that the level of user-innovation will be low (von Hippel, 1988). We had a good reason to suspect that users had a considerable need to modify and improve CD. In a telephone interview with the Director of the German Society for Dental Health and Aesthetics, we were referred to many of the product's flaws. The penetration rate of 10% also indicates that the majority of the German dentists still do not employ CD and generally obtain prosthetic teeth from dental laboratories.

Our results also support these assertions. We found a clear evidence for the existence of a considerable need on behalf of users for improving existing applications and for creating additional applications. While 48% of the respondents indicated that they desired further CD applications, 73% indicated that they had either a small or a high need for improved indications. Users were especially dissatisfied with construction suggestions provided by the software. The following comments indicate that the CD software is frequently a cause of frustration for the users:

- "15% of the suggestions are massively off-the-mark"

- "suggested restorations have to be adapted generally and considerably"
- "suggestions are frequently abysmal"

Our next task was to test the innovative potential of the CD users. Previous studies in the medical device industries and other fields have shown that if users have a need to improve a technology, they are likely to attempt developing precise improvement suggestions. According to Lüthje (2003), every fifth responding German surgeon was an innovator. We repeated this questioning approach and asked the CD users whether they had previously had an idea for an innovative medical device outside of CD. While for this question akin to Lüthje (2003), we made no efforts to test the validity of the responses, which would have certainly decreased the number of valid responses and we found that 20% asserted that they had had an idea for an innovative medical device. Of these, however, nearly half had only pursued their ideas to the sketch/outline stage. Therefore, 27 users independently reached a prototype or product stage, which amounts to 10% of the responding population. Patents were obtained for the products as diverse as a mirror with air-cooling against fogging, an information terminal or a cleaning device to extract mercury from teeth in order to create biocompatible implants. Figure 1 compares the considerable innovative potential of surgeons and dentists.

According to our understanding, the principal reason why more surgeons have innovated than dentists was that the latter are embedded in a university or a hospital environment. Most dentists operate in their own clinic, and therefore lack an access to both financial resources and the know-how available in a networked setting. The entrepreneurial nature of a dental clinic also means that less time can be devoted to experimentations.

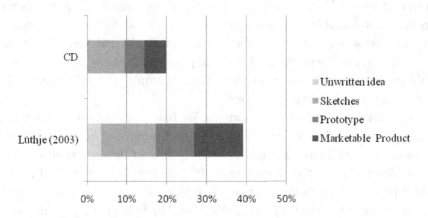

Fig. 1. The innovative potential of medical users.
Source: Authors.

To further determine the innovative potential of CD users, we found that every fourth respondent indicated that they had had a precise improvement idea for CD. Some of these ideas, such as introducing a construction toolkit in which restorations are possible from the very beginning without the systems' suggestion, using a thicker and softer material for the plastic support to increase the camera stability or to design the manual positioning of crowns in the same ways as the "inlay positioning" seem to have a considerable merit.

Having established that CD users had both the need and the potential to innovate, the next step was to determine the actual amount of modifications that they had made. We specifically asked whether users had previously, individually or through a third party, modified CD. We obtained 44% or 16% affirmative responses. As both Lüthje's (2003) and our study largely rely on the subjective assertions of users, we had to deal with the inherent response bias: Users "tend to emphasise or even overemphasise the amount of re-invention that they have accomplished" (Rogers, 2003). Through follow-up telephone calls, expert consultations and respondent interviews, we verified our results.

In this process, we found that numerous affirmative answers did not amount to modifications according to our search criteria. For example, the three users who mentioned that they had created the so-called dental-neck inlays were held not to have modified the technology, despite the fact that there was no special application for these indications, as the manufacturer specifically indicated that this was a normal process taking 2–3 minutes, which it was even taught in its training courses. Further examples of excluded modifications were the use of a third-party programme for the creation of artificial, synthetic dental provisions that the manufacturer was offering to download on its website, and the installation of new graphic cards or CPUs that could be ordered from the manufacturer or its suppliers. The creation of "extension-dental-bridges", on the other hand, was classified as a modification, as these were not specifically authorised by the manufacturer. While it was possible to create these via CD, due to the potential milling difficulties, the manufacturer did not accept responsibility for these modifications. The dentist in question, therefore, created and employed these entirely on the basis of his/her own risk-benefit assessment.

Overall, we rejected 26 declared modifications and permitted 18 modifications, leading to an overall user-modification rate of 6.6%. Given our conservative attitude in favour of including user-modifications (when we were not able to disprove the user-modification, we included it) and the response bias that these, who actually modified the technology are more likely to reply, the overall percentage of CD modifiers of the total user-population is likely to be considerably less than 5%.

Many of the modifications that we encountered were relatively minor changes that did not affect the core components of CD. Only one user tinkered with the milling device; only one allegedly modified the imaging unit and the core software remained completely untouched. While two re-wrote the system, this did not change the software as such, just their particular ability to utilise it for their purposes. Three modifications were noteworthy. Firstly, a dentist, together with a German software firm, developed a software that is yet to be commercialised, to assess the pressure on crowns and bridges and therefore to lower the risk of the indication breaking. Secondly, another user successfully created extension- and inlay-bridges, which was the most sought-after additional indication according to our questionnaire. Finally, an innovative dentist, who was frustrated with the existing method of applying powder on the teeth for the optical impression of the camera, developed a new powdering device. His PowderChamp evenly spreads the powder using air-pressure and according to a recent study, it was one of the three best (out of 15) powdering devices in the market (Anneken, 2005). Interestingly, the CD manufacturer has shown no interest in the device. Overall, especially as the PowderChamp is the only user-developed product that has been diffused/commercialised, we can summarise that the vast majority of improvements in this realm were controlled, if not made, by the manufacturer.

The final step of our analysis was to determine why this rate of user-innovation was so low. Why did not far over 90% of users modify their machines when the majority had a need to do so? As an answer to this question, we presumed that certain barriers to user-innovation exist that burden user-modifications. We, therefore, proceeded by asking respondents to assess and rank whether the following nine factors constituted barriers to user-innovation:

- CD's patent protection
- CD's copyrighted closed source code
- The adverse social stigma attached to an unauthorised product modification
- CD's hardware complexity
- CD's software complexity
- The modification prohibition in the software-license [EULA] (End-user license agreement)
- Modification would lead to a breach of warranty and a loss of authorisation as a medical device.
- It is expensive to modify CD or to have it modified by third parties.
- The effort, in terms of time, needed to improve CD

We found substantial evidence for these nine barriers. The overall median response for all the barriers was three, standing for "high barriers". The high barrier option was chosen by 53% of respondents and was by far the most frequently chosen for each individual barrier. In six of the nine areas, this option was chosen by over 50% of the users. While "hardware complexity" was held to be a high barrier by nearly two-thirds of respondents, three-fourths chose this option for "software

Table 1. Barriers to CD user-innovation.

	Patents	Copyright/ Closed source code	Hacking	Hardware complexity	Software complexity	EULA	Warranty/ Medical device authorization	Expense	Time & effort	Total
N =	209	206	199	189	184	201	209	205	210	1812
No barrier	29%	18%	25%	13%	8%	23%	20%	20%	15%	19%
Low barrier	15%	15%	14%	11%	8%	16%	8%	9%	10%	12%
Med. barrier	15%	15%	19%	12%	10%	19%	16%	21%	20%	16%
High barrier	41%	52%	43%	64%	73%	42%	57%	51%	55%	53%
Average	1.7	2.0	1.8	2.3	2.5	1.8	2.1	2.0	2.14	2.0
Median	2	3	2	3	3	2	3	3	3	3

0 = No barrier; 1 = Low barrier; 2 = Medium barrier; 3 = High barrier.

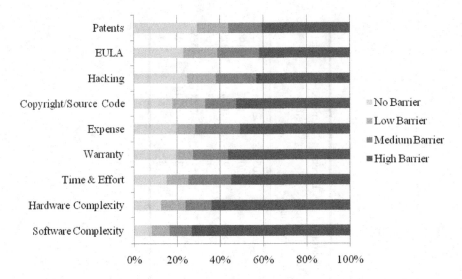

Fig. 2. Barriers to CD user-innovation.
Source: Authors.

complexity". Around 70% of all the users classified the nine factors as medium or high barriers to user-innovation. As a caveat, we do need to point out that these data merely refers to user perceptions of an exclusionary mechanism.

Table 1 and Fig. 2 summarise our results. We ranked the different barriers and found substantial differences among them. While hardware and especially software complexity were considered to be very high barriers, patents, hacking and restrictive software licenses received a far less support. Nevertheless, with medians of 2 and averages of 1.67, 1.8 and 1.8, respectively, even these were still considered to be moderate barriers. The distinction, nevertheless, touches upon the important difference between *ex-ante* and ex-post consequences of the barriers. Complexity and the time and effort needed to innovate operate *ex-ante* or before the user can improve the technology. Patent infringement, breach of a software license or the social stigmatisation of hacking on the other hand operate mainly after the user has already made the improvement. While they can prevent the diffusion of the development, if detected, they previously serve merely as a psychological barrier, which is far easier to overcome than a tangible one.

Manufacturer-generated barriers to user-innovation in CD

The essential question for our purposes at this point is to consider which of these barriers were imposed by the manufacturer. Through our research, we found various indications for the assertion that the manufacturer was striving for what one user termed, "complete control over the product and its uses". The manufacturer apparently exerted control not only over how the product is specifically used but also over all training courses and to some extent, over the most important publications in this realm.

In order to see what weapons are used in this process, let us begin with the legal barriers. Patents are legal rights that allow their holders to exclude anyone else from making, using and selling a specific invention for a set period. Whether or not patents will be used to pursue infringers will depend on the strategic situation in question. For IBM, for example, it made commercial sense to grant the open-source movement free access to some of its software patents, as this increased the demand for its hardware and services (Benkler, 2002). For the CD manufacturer, it made commercial sense to closely guard its IP. According to our interviews, the manufacturer held 135 CD patents globally, which apparently was a key reason why competitors had so far been marginalised in this area. While patents showed up as a moderate barrier, this was largely due to the limited intention of the dentists to commercialise their modifications.

End-user license agreements (EULAs) regulate the rights to use a specific software. EULAs have contractually prohibited millions of users from amending, reverse engineering, reproducing or modifying all types of non-open-source software programs (Harper and Udupa, 2004). The CD EULA contains a standard non-modification clause and serves as a warning that user tinkering is prohibited. The moderate deterrent value that we found in our study may again be due to the marginal importance of diffusion.

While the purely legal barriers did not amount to the main obstacles for user-modifications, the combination of copyright protection and the closing of the source code constituted a more significant barrier. Fifty-two per cent of the respondents rated it as a high barrier and two-thirds as a moderate or high obstacle. The closing of a software's source code is a purposive strategic decision against the direct user-participation in the software improvement process. Unlike the threat of patent or EULA infringement, this barrier is a part of the product architecture and therefore operates *ex-ante*. Anti-circumvention laws such as the US Digital Millennium Copyright Act or the European Union's Copyright Directive reinforce such product control. In order to modify the closed software, users have to first reverse-engineer the code. The difficulty, expense and time needed to reverse-engineer frequently make modification attempts unattractive (Samuelson and Scotchmer, 2002). While not making user-modifications impossible *per se*, the decision to close certainly drives up the costs of such tinkerings substantially. Innovation with respect to closed software will, according to Dahlander (2006, p. 20), thus takes place "within the boundaries of the firm or is supplemented by the contractual relations with suppliers and consultants. Customers cannot further improve the software, but merely can send suggestions to the firm with new functionalities and demand or hope so that the vendors fix the bugs".

Numerous CD users responded that it was both too time-consuming and expensive to modify the technology in question. For the medium and high barrier options, 75% and 72% users chose these options, respectively. The non-modularity of the product, again a purposive choice by the manufacturer, contributed to this difficulty. Opening the source-code could have a tremendous effect on user-innovation in this field. We, therefore, inquired how many CD users believed that they could improve the CD software, if it were an open-source. Forty-one per cent of the respondents or 108 users answered affirmatively. To test the coherence of these responses, we compared the percentage of their verified modifications with those who indicated that they could improve an OSS version of the CD. We found that while 10% of the OSS enthusiasts had previously modified CD, this applied to only 4% of users, those who responded that they lacked the requisite skills. This lends support to the view that closing the source code can stifle potentially valuable user-modifications.

Another mechanism to discourage user-experimentation is the introduction of a strict warranty. In our survey, 57% of the respondents chose the high-barrier option. In this case, the CD manufacturer attached seals to the PC slide model, which stated that "if the seal is broken, all warranty claims against the PC slide-in module are invalid". In its CD hardware compliance requirements, the manufacturer further states that:

> "The [CD] acquisition unit hardware supplied by [the manufacturer] has been specifically selected and matched for our software applications. This combination gives optimal performance of the [CD], along with meeting the strict requirements of the FDA .... The use of any hardware, other than what is supplied by [the manufacturer] will violate the compliance of the [CD] and compromise its safety and efficacy. The use of non-approved hardware will void the warranty and support from [the manufacturer]"

The principal reason why such words may discourage some users is the capital-intensity of the device. Clearly, a user is less likely to ignore the terms of the warranty of an expensive device, which he/she the generally cannot service himself/herself than in the case of a cheap gadget. While this ex-post barrier may not prevent users from tinkering, it does increase the costs of doing so. As an anecdotal evidence, a professional photographer explained that the fear of breaching the warranty for his expensive Stylus Photo printer led him to only use independently filled cartridges, once the warranty had expired (Seattle Times, 2006).

A related risk for a tinkering physician is that modifications are likely to lead to the loss of CD as an authorised medical-device. An unauthorised modification could lead to a malpractice claim, in case a patient suffers from resulting damage. Various courts, especially in the United States, have held doctors liable for harms resulting from experimenting with unproven procedures or technologies (Curran, 1984). Although the legal environment is less virulent in Germany, an unauthorised modification will still be a risky undertaking for a doctor. The way that a manufacturer can reinforce this fear is by systematically investigating the use of its machine and reporting any cases of malpractice. Through such patrolling, a similar culture of fear could evolve that has been reported by some seed-farmers (Bauer, 2005).

This closely relates to the weapon of social condemnation. The term, "hacker", for example, originally referred "to computer enthusiasts ... who through clever programming, pushed a system to its highest possible level of performance" (Wellen and Reiter, 1998) and to "heroes of the computer revolution" (Levy, 1984). In the early 1980s, corporations, governments and the media "began to bastardise the term to include criminal conduct" (Wellen and Reiter, 1998). Today, the term "hacker" generally has a negative connotation attached to it, ignoring any positive aspects of such behaviour. As a result, many users without permission to tinker, especially underground or pirate user-innovators, have been stigmatised as criminals (Mollick, 2004). Hope (2004, p. 221) adds that this "fear, uncertainty and doubt" strategy was purposefully adopted by some software companies, most notably the Microsoft, to combat the advance of an open-source software. While in our survey, this social condemnation of hacking was only of moderate importance, we encountered some anecdotal evidences in subsequent interviews that the manufacturer was combating innovative users whom they saw as a threat to its business model. One user, for example, purchased old machines, repaired and modified these for re-sale. The user complained of pressure being exerted on him/her, for example, in the form of journals being asked not to publish his/her articles.

We wanted to test the sensibility of this strategy in respect of Mollick's (2004) and Flowers's (2008) findings that outlaw users are frequently innovative and of potential value to the manufacturer. While the CD manufacturer collected user-feedback, many respondents seemed to be disillusioned by this inadequate attempt to appreciate user ideas and developments. This is unfortunate, as 76% of the users believed that they could further improve the technology in case of a closer co-operation with the manufacturer. While this certainly amounts to an overestimation of their personal abilities and resources, it does indicate a desire to help improve CD. In our survey, we further found that 30% of the users were dissatisfied with the current interaction with the manufacturer. This is not only a considerable amount in itself but also of significance from a user-innovation perspective. The users with the highest need tend to have the highest incentive to modify. Dissatisfied users should,

Table 2. The innovative potential of dissatisfied users.

	General population	Dissatisfied users	Δ[a]	Δ in %
N =	274	74		
Believe that they could modify	76%	88%	+12	+16%
% Self-assessed high ability to modify CD	12%	19%	+7	+60%
Average self-assessed skill-set	1.25	1.33	+0.1	+6%
% modification (self-declared)	16%	18%	+2	+12%
% modification (verified)	7%	13%	+6	+102%

[a]Percentages refer to exact figures.

therefore, be the more valuable co-operation partners. Our findings strongly support this assertion. We found that dissatisfied users modified 102% more than the satisfied ones. Table 2 summarises these results. In our triangulation efforts, we encountered two further dissatisfied users, who had used the software in novel ways. On this basis, it seems sensible for the manufacturer to re-think its approach of dealing with highly dissatisfied and outlaw users.

At this stage, a few words about purposefully imposed technological barriers seem to be appropriate. Our survey found that technological complexity of both hardware and software was the main obstacle to user-modifications. An overwhelming 76% and 83% of the respondents chose these barriers as moderate or high barrier. There is an inherent component of complexity that seems to operate largely independent of the manufacturer. In our survey, we found that 52% of the respondents believed that the current version of CD was more difficult to modify than its previous version. Twenty-eight per cent believed that it had become easier. While unfortunately, we were not able to determine why respondents believed that the difficulties of modifying CD had largely increased, there seems to be a twin cause. Apart from the increase in technological sophistications, which increases the skills that users need to improve the product in question, the manufacturer clearly chose, in line with Douthwaite's previously mentioned view, to make user-tinkering as difficult as possible. Apart from the closed source code and the non-modularity of the product, the interaction between the software and the hardware, over time, has been steadily optimised to make it difficult for the outsiders to enhance the system. Furthermore, the CD hardware is sealed and is not supposed to be opened by users, as clearly voiced by the instruction manual: "Opening of the unit with tools is strictly prohibited". In this manner, as some users even openly complained, the manufacturer can largely keep the whole workings of the technology "secret".

We have now addressed some of the legal, economic, technological and social weapons that were used in the case of CD to discourage user-modifications. A manufacturer seems to be able to preserve his role as the innovator in a field by making tinkering excessively difficult. The statement of one CD user summarises the *status quo* succinctly: "the effort [and cost] involved in tinkering has to lead to a net-gain, which it currently does not". Lüthje and Herstatt's (2004) question whether manufacturers can actively change users' cost and benefit expectations with respect to innovations, therefore, clearly receives an affirmative response.

Further Freedom Fighters

When it comes to preventing user-innovation, the field of computerised dentistry is no outlier. To indicate the significance of our findings, we will now give numerous examples of incumbents attempting to control and limit user-innovation in different ways.

To begin with, let us return to the printer industry. In order to protect their business models, printer manufacturers had to find a way of preventing aftermarket cartridge re-filling. Their solution was to introduce microchips into their cartridges. These intelligent chips can detect non-authorised cartridges and react accordingly by substantially decreasing the print quality or impairing/halting any further operations of the printer (Spring, 2003; Knapp, 2003). Such sophisticated cartridge chips have become a common practice and make it progressively difficult for aftermarket users to reverse-engineer them. Wagstaff (2002) commented that printer companies are "building the cartridge cases like Fort Knox, leaving only the brave to fiddle with them". Golden (2002, p. 36) concurred by saying that printer manufacturers are attempting to "erect barrier-after-barrier in an effort to severely limit or even completely eliminate the aftermarket".

While the anti-trust law could prevent such practices, it will not apply if a company lacks the market power to fall under anti-trust regulations. In *Lexmark v Static Control*, a US Court of Appeal even legitimised such corporate behaviour by stating that "manufacturers of inter-operable devices such as computers and softwares, game consoles and video games, printers and toner cartridges or automobiles and replacement parts may employ a security system to bar the use of unauthorised components". Strong patents and some market power back up the printer companies' arsenal to constrain user-innovation. A recent law suit even alleges that Hewlett-Packard paid a large office retail store, a multi-million dollar sum to stop selling HP compatible off-brand ink cartridges (Gaylord, 2008). Although the forecasts are that the aftermarket's share of the North American ink cartridge sales will continue to grow (Lyra Research, 2005), such hurdles have forced the aftermarket to professionalise its tinkering, as the wide variety of cartridge re-filling companies that have been formed illustrates. While in the late 1990s, "a single technician with minimal equipment [and resources] could conduct the entire [reverse engineering and adjusting] operation", today highly trained "team[s] of specialists from a half-a-dozen disciplines, combined with an automated, robotic production machinery" require many months. Developing an appropriate aftermarket solution for the HP 4100 Printer, for example, took a total of 13 months (Golden, 2002).

In the field of agriculture, seed-manufacturers such as Monsanto have largely succeeded in suppressing user-innovation through the ruthless employment of technological, contractual, legal and even social barriers (Braun and Herstatt, 2007). Restrictive contractual licenses prohibit farmers from saving seeds. Through the inclusion of marker genes to identify their proprietary seeds, the vigorous enforcement of their patent rights has reached another dimension. In *Monsanto v Ralph* (2004), for example, a farmer who commercialised proprietary farm-saved seeds was obliged to pay $3 million as a reasonable royalty fee. Through hybridisation of seeds, they have furthermore made seed-saving, the pre-requisite for farmer seed-breeding, futile. By rewarding farmers, who report those who they believe are violating Monsanto's rights, a social barrier of fear was created (Bauer, 2005). For Monsanto, this strategy of offering users an excellent product while excluding them from the innovation function has paid off handsomely. Each of the last five years has seen substantial increases in both revenue and net income and their shares have soared from around $7 in early 2004 to over a $100 at the end of 2007 (Hindo, 2007; Monsanto Annual Report, 2007).

The automobile industry has also been quite hostile when it comes to innovative users. In the industry's early days, manufacturers like Ford used strong warranties and even pressure on dealers to warn users not to modify their cars (Shah and Tripsas, 2004). Given that the gross profit margin of selling new cars has been falling and the average age of cars increasing, the incentives to make it as difficult as possible for outsiders to modify their vehicles and thus, generate business for their own aftermarket services are obvious (Englezos, 2006). Currently, the vast majority (around 75%–80%) of this over US$200 billion industry is usurped by independent repair shops (Mello Jr., 2005). Many car components, therefore, contain embedded chips and interdependent components. The motors of many cars are now built in one non-modular unit that makes user-modification challenging. The bonnet of the Audi A2, for example, is completely closed as an access to the inner workings of the car is only intended by authorised Audi dealers. If users do open their hood, their warranty will terminate (Lockton, 2005). The increased computerisation of cars has further made various types of repairs dependent on an access to diagnostic codes, which are already contained in emission control systems, anti-lock brakes, climate control, airbags and other safety systems. A 2002 membership survey by the US Automotive Association, which represents over 15,000 independent repair shop owners, found that 10% of cars could not be repaired due to the unavailable codes and lack of adequate code scanning equipment (Pickler, 2002). A 2006 AAIA survey of 1000 independent repair shops supported this finding that a large amount of codes and necessary tools were missing, especially with respect to foreign cars (AAIA, 2006). Currently, in the United States, a voluntary scheme is in place that gives independents, but not users themselves, an access to the information and diagnostic tools necessary to service vehicles, in return for a license fee and a confidentiality promise (Mello Jr., 2005). While the leading car manufacturers have not managed to win the battle for the aftermarket, they have made user-innovation in this field increasingly dependent on professional commitment.

While there are numerous other examples of incumbents burdening user-innovations, we will conclude this section by briefly focusing on Apple. Apart from its design eloquence, the essence of Apple's successful business model is its independent and "hermetically sealed [product development] system" (Penenberg, 2007). Apple continues to refuse to license its operating system and has, for instance, included special security chips in its Macs to prevent people from loading it onto non-Apple machines (Singer, 2005). While the closed product architectures for its popular iPod and iPhone gadgets has enabled Apple to protect its know-how from competitors and to reap high profits by selling incrementally enhanced products, Tapscott and Williams (2006, p. 134) raise the critical question that leads us into the last section of this paper: "Is a business model that locks in customers and discourages user-innovation genuinely sustainable?"

Implications: Can Exclusion be Sustainable?

We have established that suppressing user-innovation is currently a widespread corporate activity. In line of the democratisation argument, the question is raised whether corporations are fighting a battle that they will inevitably lose? The music industry comes to mind as a field where companies, under the representation of the notorious Recording Industry Association of America (RIAA), are losing out to users. Digitisation and the Internet have enabled these companies to copy music to previously unimaginable extents. The aggressive prosecution of thousands of users and the shutting down of one P2P server after the other, have not rescued the music industry's bottom line (Tang, 2005). Its old business model is simply no longer sustainable and new ways of commercialisation music have to be found. In this light, we want to end this paper by summing up our thoughts under which conditions a strategy of excluding users from the R&D function is likely to remain profitable in the long-run.

Lego, on the other hand, has realised that allowing and even encouraging users to modify their products can be a more promising strategy than exclusion. When the Danish company introduced its programmable Lego robot, the Mindstorm, a group of hackers deciphered its proprietary code, posted it on the Internet and subsequently wrote various software applications that extended the robot's capabilities (Keegan, 2001). They even created a new operating system, LegOS. Lego could have sued the hackers for circumventing a copy-right protection device, but chose to include their developments in future Mindstorm versions and to encourage their tinkering via tool kits (Labrador, 2002). More recently, Lego has continued to integrate users into its R&D operations by launching the Lego Factory, in which an interactive software allows and assists users to design and share their own customised Lego models. While this operation offered users a unique service, they still hacked the platform in order to improve the brick allocation. Although this hack substantially lowered the average price tags for the custom models, Lego still appreciated such user involvement, as it increased the popularity of its platform (Terdiman, 2005).

Sony, a company that has traditionally prosecuted outlaw users, such as mod developers or Playstation hackers, with all the available means had a similar opportunity in respect of its Aibo pet. One user reverse-engineered the digitally encrypted proprietary code of this robotic pet and thereby *inter alia* taught the Aibo to dance jazz (Lessig, 2004). Sony originally threatened to sue the user and demanded the removal of the programs to maintain control over their technology (Bovens, 2005). Due to an uproar in the Aibo community, Sony now permits the distribution of the experimental software and has even integrated the hacker's developments into the future Aibo versions (Labrador, 2002).

What do these anecdotes teach us about the conditions under which exclusion is likely to be profitable? The authors of *Wikinomics* acknowledge that currently "product hackers are still a small minority of [Sony's or Apple's] customers, and [that] there is little evidence yet that products hacks and home-brew applications are leaking into the mainstream". Nevertheless, they believe that "mass collaboration ... will eventually displace the traditional corporate structures as the economy's primary engine of wealth creation" (Tapscott and Williams, 2006, p. 136). They base their confidence on the changes in the business environment. They explain that mass user-collaboration depends on three main conditions: low participation costs for users, modular tasks and low costs of assembling the pieces into a final end product. The existence of an information commons (von Hippel and von Krogh, 2006) and a sizable group of users with the requisite skills, as well as the absence of IP barriers and effective technological restraint mechanisms are also critical for mass user-innovation to work.

If manufacturers are able to successfully impair one of these components, an exclusion strategy may operate effectively. The initial question is always whether users have a need to innovate. If a market consists of relatively homogenous needs, like the western seed markets where climate conditions are comparatively similar and a manufacturer, like Monsanto, can satisfy this need, an exclusion is likely to be feasible. If the market need is heterogeneous, like seed-markets in developing countries and the manufacturer can only partly meet this need, an exclusion is likely to be much more difficult.

The next question concerns the ability to modify and the costs thereof. This depends on the existence of an information commons from which users can freely obtain some of the innovation perquisites such as tools or know-how. While such commons exists in the field of software in other areas, know-how is not widely available. While there are efforts to find cures for tropical diseases in open-source modes (Maurer et al., 2004), these have so far not lead to any successful results. This is largely due to the absence of a critical mass of skilled users, who have an access to the necessary infrastructure and supplies in the field of biotechnology (Hope, 2004). A further critical difference is the much higher cost and difficulty of both producing and diffusion biotechnology products. While in the field of software possessing, a computer and a fast Internet connection can be sufficient to develop and diffuse a new program, the pre-requisites to innovate are currently much higher for biotechnology products. Currently, around 90% of R&D is still performed internally (Tapscott and Williams, 2006) indicating that the vast majority of the cutting edge know-how is not freely available to interested users. If all the leading manufacturers in a field decide to keep the innovation pre-requisites closed, an exclusion could be sustainable. This is, obviously, subject to some extent to public outcry and subsequent changes in the governmental policy.

The type of market and product involved will also be of critical importance. In a recent study in Germany, Schultz et al. (2007) found substantial differences in user involvement between the innovation processes of the test-engineering and textile-engineering industries. While the former had a high extent of user-involvement, the latter refused to collaborate with lead-users. The main difference between these two sets of firms was that the textile market is highly saturated in which the main drivers of competition are costs, while in the testing market, the critical success factors are quality and product differentiation, a task in which customers can contribute substantially.

The extent of a company's market control and the ability to erect barriers to entry are also important influential factors. Monsanto or Microsoft can afford to close their systems due to their requisite market powers. Both of them have erected high barriers to entry in the form of high switching costs and obstacles to inter-operability and proprietary genes, respectively. The success of Apple's iPod and iPhone has also largely resulted from the absence of competition to its innovative products (Penenberg, 2007). As competitors with more open platforms are, however, preparing their attack, Sprint, for example, has a touch-screen phone that runs thousands of third-party applications; Apple was forced to react and thus successfully released a third-party software development kit.

The strength of IP rights, especially patents, and the ability to enforce these patent rights will also matter. Patents tend to be of relatively little actual importance when an enforcement is difficult. Enforcement involves the probability of detection and the adverse consequences of enforcement. While software patents may be able to legally prevent some forms of open-source programming, the high costs associated with mass enforcement against thousands of users and the resulting bad publicity will make such course of action unattractive for incumbents. If, however, patent rights are strong and enforcement costs are relatively low, as is the case in pharmaceuticals industry, patents can be a nearly insurmountable barrier (Mansfield, 1986; Cohen et al., 2000).

The extent of market regulation can also be a decisive factor in deciding whether users can be excluded from certain innovation tasks. While very little, if any regulatory hurdles need to be fulfilled before diffusing software, new medical devices and pharmaceutical products generally require previous regulatory approval. Various studies have confirmed that the long duration and expense of such approval processes, while necessary to safeguard patients can constitute a substantial "barrier to innovation" (BMBF, 2005). In a recent study of the German medical device industry, 95.2% of the respondents found that such laws and regulations were endangering the industry's growth (Spectaris, 2004).

The type of technology also has a substantial effect on the sensibility of an exclusion strategy. If you can implement effective, technological protection mechanisms around your product making circumvention impossible or very costly and time-intensive, an exclusion may be sensible. The failure of digital-rights protection (DRMs) in the music industry (Tang, 2005) indicates that developing tinker-resistant products is frequently a challenging endeavour. Apple and the car and printer industries are still largely in the process of trying, with mixed success, to make their products tinker-resistant. Genetically modified seeds and hybrids, on the other hand, can prevent user-modifications. With respect to software, closing the source code and embedding the security tools directly in the software's binary code can make tinkering very difficult (Dahl, 2007). Lie (2003) further argues that through a combination of software and hardware code, tinkering can be made completely tamper-resistant. Such "XOM code" can only be executed, but not read or modified. Hardware components itself can be physically "encapsulated to make non-destructive disassembly [and copy] almost impossible" (Pooley, 2001, Secs. 5–25).

To conclude this paper, companies will always have to evaluate which problems are best solved via an open and which via a closed approach. It is certainly questionable whether ignoring and combating the innovative potential of users is likely to be a successful long-term strategy. This does not imply that abandoning the traditional structures of vertical integration is likely to be the holy grail of corporate success. Instead of combining the two approaches, depending on the requisite circumstances, it seems a lot more promising strategy. Companies should not automatically assume that it is better to keep critical IP and know-how secret. They should internally analyse when it makes sense to open-product development to users. It will probably never make sense for Coca Cola to reveal its secret formula. Continuing to protect innovative seeds, certain software kernels or other proprietary platform components are also likely to remain commercially sensible. Apple, IBM and Lego seem to have realised, however, that opening up parts of their platforms can have significant benefits in terms of improving their corporate image, enhancing the popularity of their products and in the long-term increasing market share and profits. In this light, we concur with Schultz *et al.*, (2007), who emphasise that research into companies proactively refusing to integrate users may amount to a fruitful venue for future studies. Scrutinising the other side of the user-innovation and its relationship with the democratisation perspective is essential for anyone who is interested in fully understanding the future of new product development.

References

AAIA (2006). Car company vehicle service information & tools: An evaluation of their availability to independent repair shops. Automotive Aftermarket Industry Association.

Anneken, T (2005). *Vergleich Verschiedener Puder- und Lack Systeme zur Optischen Kavitätenvermessung bei Cerec 3D*. Hamburg: Medizinische Fakultät, Universität Hamburg.

Arora, A, A Fosfuri and A Gambardella (2002). *Markets for Technology: The Economics of Innovation and Corporate Strategy*. London: MIT Press.

Bauer, A (2005). Eine ökonomische Katastrophe. *Umweltnachrichten*, 102 (December 2005).

Benkler, Y (2002). Intellectual property and the organization of information production. *International Review of Law and Economics*, 22(1), 81–107.

Berinato, S (2003). The future of security. *CIO*, 17(6), 70–76.

Berners-Lee, T (2000). *Weaving the Web: The Past, Present and Future of the World Wide Web by its Inventor*. London: Texere.

Biemans, W (1991). User and third-party involvement in developing medical equipment innovations. *Technovation*, 11(3), 163–182.

BMBF (2005). *Studie zur Situation der Medizintechnik in Deutschland im Internationalen Vergleich*. Berlin: Bundesministerium für Bildung und Forschung.

Bovens, A. (2005). Closed architectures for content distribution. *Japan Media Review*, (12 February 2005).

Braun, V and C Herstatt (2007). Barriers to user innovation: Moving towards a paradigm of 'licence to innovate'? *International Journal of Technology, Policy and Management*, 7(3), 292–303.

BSA (Business Software Alliance) and IDC (International Data Corporation) (2006). *Third Annual BSA and IDC Global Software Piracy Study*, May 2006.

Chesbrough, H (2003). *Open Innovation — The New Imperative for Creating and Profiting from Technology*. Boston: Harvard Business School Press.

Chesbrough, H and A Kardon Crowther (2006). Beyond high tech: Early adopters of open innovation. *Research Policy*, 36(3), 229–236.

Cohen, W, R Nelson and J Walsh (2000). Protecting their intellectual assets: Appropriability conditions and why US manufacturing firms patent (or not). Working Paper 7552. Cambridge, MA: National Bureau of Economic Research.

Curran, W (1984). The unwanted suitor: Law and the use of health care technology. In *Machine at the Bedside*, S Reiser and M Anbar (eds.), 119–133. Cambridge: Cambridge University Press.

Dahl, D (2007). Case studies: A hacker in India hijacked his website design and was making good money selling it. Inc., December 2007, 77–80.

Dahlander, L (2006). Managing beyond firm boundaries: Leveraging user innovation networks. Department of Technology Management and Economics, Gothenburg: Chalmers University of Technology.

Dahlander, L and M Magnusson (2005). Relationships between open source software companies and communities: Observations from Nordic firms. *Research Policy*, 34(4), 481–493.

Douthwaite, B (2002). *Enabling Innovation: A Practical Guide to Understanding and Fostering Technological Change*. London: Zed Books.

Engardio, P and B Einhorn (2005). Outsourcing innovation, *Business Week* (3925), 84.

Englezos, P (2006). *A Cross-Industry Analysis and Framework of Aftermarket Products and Services*. Cambridge, MA. Massachusetts Institute of Technology.

Enos, J (1962). *Petroleum Progress and Profits: A History of Process Innovation*. Cambridge, MA: MIT Press.

Feller, J and B Fitzgerald (2002). *Understanding Open-Source Software Development*. Boston: Addison Wesley.

Flowers, S (2008). Harnessing the hackers: The emergence and exploitation of outlaw innovation. *Research Policy*, 37(2), 177–193.

Focus (2006). iPods sollten Sprengstoff zünden. *Focus* (11 August 2006).

Gassmann, O (2006). Opening up the innovation process: Towards an agenda. *R&D Management*, 36(3), 223–228.

Gaylord, C (2008). Why printer ink is so expensive. *CS Monitor*, January 9, 2008.

Golden, C (2002). Toner cartridge computer chip usage and the impact on the aftermarket. *Recharger Magazine* (December 2002), pp. 36–46.

Graaf, B, M Lormans and H Toetenel (2003). Embedded software engineering: The state of the practice, *IEEE Software* (November/December 2003), 61–69.

Granstrand, O (1999). *The Economics and Management of Intellectual Property: Towards Intellectual Capitalism*. Cheltenham, U.K.: Edward Elgar.

Greenberg, A (2008). China clamps down on internet video. *Forbes* (1 March 2008).

Guinnessy, P (1997). Tooth delay banished. *New Scientist*, 2090 (12 July 1997).

Harper, B and V Udupa (2004). Drafting electronic software licenses to prevent reverse engineering. *E-Commerce Law Report*, 6(2), 18.

Henkel, J and E von Hippel (2005). Welfare implications of user innovation. *J. Technology Transfer*, 30(1–2), 73–87.

Herstatt, C and E von Hippel (1992). From experience: Developing new product concepts via the lead user method. *Journal of Product Innovation Management*, 9(3), 213–222.

Heuer, S (2006). Die vertriebs-täter. *Brand Eins*, 8(7), 26–34.

Hienerth, C (2006). The commercialization of user-innovations: The development of the rodeo kayak industry. *R&D Management*, 36(3), 273–294.

Hindo, B (2007). Monsanto: Winning the ground war: How the company turned the tide in the battle over genetically modified crops. *Business Week* (17 April 2007).

Holt, K (1988). The role of the user in product innovation. *Technovation*, 7, 249–258.

Hope, J (2004). *Open-Source Biotechnology*. Canberra: The Australian National University.

Howells, J (1999). Research and technology outsourcing. *Technology Analysis & Strategic Management*, 11(1), 17–29.

Industrial Research Institute (IRI) (2007). Industrial research institute's R&D trends forecast for 2007. *Research-Technology Management*, 50(1), January–February 2007, 17–20.

Joy, B (2000). Why the future does not need us? *Wired*, 238–262 (August 2004).

Keegan, P (2001). Lego: Intellectual property is not a toy. *Business 2.0* (September 2001).

Klein, B (1996). Market power in aftermarkets. *Managerial and Decision Economics*, 17(2), 143–164.

Kline, R and T Pinch (1994). *Taking the Black Box Off its Wheels: The Social Construction of the Car in the Rural United States*, Berlin: WZB Wissenschaftszentrum Berlin für Sozialforschung.

Knapp, L (2003). Printer fails after run-in with generic ink cartridge (17 May 2003). *The Seattle Times*, Seattle.

Kurzweil, R (1999). *The Age of Spiritual Machines*. New York: Penguin Books.

Labrador, D (2002). Teaching robot dogs new tricks (21 January 2002). *Scientific American*.

Lessig, L (1999). *Code and Other Laws of Cyberspace*. New York: Basic Books.

Lessig, A (2004). *Free Culture*. New York: The Penguin Press.

Lettl, C, C Herstatt and HG Gemuenden (2006). User's contributions to radical innovation: Evidence from four cases in the field of medical equipment technology. *R&D Management*, 36(3), 251–272.

Levy, S (1984). *Hackers: Heroes of the Computer Revolution*. London: Penguin.

Lie, D (2003). *Architectural Support for Copy and Tamper-Resistant Software*. Stanford University: Department of Electrical Engineering.

Lilien, G, P Morrison, K Searls, M Sonnack and E von Hippel (2002). Performance assessment of the lead user generation process for new product development. *Management Science*, 48(8), 1042–1059.

Lockton, D (2005). Architectures of control in consumer product design. Judge Institute of Management, Cambridge: University of Cambridge.

Lotz, P (1992). *Demand Side Effects on Product Innovation: The Case of Medical Devices*. Copenhagen: Copenhagen Business School.

Lüthje, C (1999). Kundenorientierung im Innovationsprozeß: Eine Untersuchung zur Customers–Hersteller–Interaktion auf Konsumgutermärkten. Munich, Ludwig–Maximillians–Universität München.

Lüthje, C (2003). Customers as co-inventors: An empirical analysis of the antecedents of customer-driven innovations in the field of medical equipment, *Proc. 32nd EMAC Conference*, Glasgow.

Lüthje, C and C Herstatt (2004). The lead user method: An outline of empirical Findings and issues for future research. *R&D Management*, 34(5), 553–568.

Lüthje, C, C Herstatt and E von Hippel (2005). The dominant role of "local" information in user innovation: The case of mountain biking. *Research Policy*, 34(6), 951–965.

Lyra Research (2005). *Think Ink: The 2005 US Ink Jet Cartridge User Survey*. Newtonville, MA: Lyra Research.

Mansfield, E (1986). Patents and innovation: An empirical study. *Management Science*, 32(2), 173–181.

Maurer, S, A Rai and A Sali (2004). Finding cures for tropical diseases: Is Open Source an answer? *Biotechnology: Essays from its Heartland*, L. Yarris, Lawrence Berkeley Laboratory, 33–37.

Mello Jr, J (2005). Bill would force car makers to share computer code. (5 May 2005). *TechNewsWorld*.

Mollick, E, (2004). *Innovations from the Underground: Towards a Theory of Parasitic Innovation*. Cambridge: MIT (133).

Monsanto (2007). *Monsanto Annual Report 2007*, St. Louis.

Morrison, P, J Roberts and D Midgley (2004). The nature of lead users and measurement of leading-edge status. *Research Policy*, 33(2), 351–362.

Mossberger, K, C Tolbert and M Stansbury (2003). *Virtual Inequality*. Washington, D.C.: Georgetown University Press.

Nagel, R (1993). *Lead User Innovationen -Entwicklungskooperationen am Beispiel der Industrie elekronischer Leiterplatten*. Wiesbaden: Deutscher Universitäts-Verlag.

Neff, G and D Stark (2004). Permanently beta. *Society Online: The Internet in Context*, J Howard (ed.), pp. 173–188. Thousand Oaks: Sage.

Olson, E and G Bakke (2001). Implementing the lead-user method in a high-technology firm: A longitudinal study of intentions vs. actions. *Journal of Product Innovation Management*, 18(2), 388–395.

Pallesen, U and J van Dijken (2000). An 8-year evaluation of sintered ceramic and glass ceramic inlays processed by the Cerec CAD/CAM system. *European Journal of Oral Science*, 108, 239–246.

Pearson, I (2005). Im Jahre 2030 leben wir alle in digitalen Blasen (21 January 2005). *VDI Nachrichten*, Ipswich, England.

Penenberg, A (2007). All eyes on Apple. *Fast Company*, 121, 83–87; 133–136. December 2007.

Peritz, R (2002). Antitrust policy and aggressive business strategy: A historical perspective on understanding commercial purposes and effects. *Journal of Public Policy & Marketing*, 21(2), 237–242.

Pickler, N (2002). Mechanics struggle with diagnostics. (24 June 2002). *Associated Press*.

Pooley, J (2001). *Trade Secrets*. New York: Law Journal Press.

Prahalad, C and V Ramaswamy (2004). *The Future of Competition: Co-creating Unique Value with Customers*. Boston, MA: Harvard Business School Press.

Ramírez, R (1999). Value co-production: Intellectual origins and implications for practice. *Strategic Management Journal*, 20, 49–65.

Reiss, B (2006). Clinical results of Cerec inlays in a dental practice over a period of 18 years. *International Journal of Computerized Dentistry*, 9(1), 11–21.

Rice, R and E Rogers (1980). Reinvention in the innovation process. *Knowledge*, 1, 499–514.

Riggs, W and E von Hippel (1994). Incentives to innovate and the sources of innovation: The case of scientific instruments. *Research Policy*, 23(4), 459–470.

Rogers, E (2003). *Diffusion of Innovations*. New York: Free Press.

Samuelson, P and S Scotchmer (2002). The law and economics of reverse engineering. *The Yale Law Journal*, 100(6), 1575–1663.

Schultz, C, M Nolting, K Dautzenberg, G Müller-Seitz and G Reger (2007). Lead using or Lead refusing? An exploratory examination of open innovation activities by lead users in mechanical engineering. *Proc. of the 2007 IEEE International Conference on Industrial Engineering and Engineering Management*, 2–4 December 2007.

Seattle Times (2006). Printer fails after run-in with generic ink. (6 June 2006). Response to *Seattle Times Article*.

Shah, S and M Tripsas (2004). When do user-innovators start firms? Towards a theory of user entrepreneurship. Boston, MA: Harvard Business School Working Paper.

Shapiro, AL (2000). *The Control Revolution — How the Internet is Putting Individuals in Charge and Changing the World We Know*. New York: NY, Public Affairs.

Shaw, B (1985). The role of the interaction between the user and the manufacturer in medical equipment innovation. *R&D Management*, 15(4), 283–292.

Singer, M (2005). Coders worried by DRM chip in Apple's Intel box (5 August 2005). *CNET News.com*.

Spanbauer, S (2003). Product activation gains ground. *PC World*, 21(6), 36.

Spectaris (2004). *Die deutsche Medizintechnik 2004*, Spectaris.

Spring, T (2003). Why do ink cartridges cost so much? *PC World* (28 August 2003).

Tang, P (2005). Digital copyright and the 'new' controversy: Is the law moulding technology and innovation? *Research Policy*, 34(6), 852–871.

Tapscott, D and A Williams (2006). *Wikinomics: How Mass Collaboration Changes Everything*. New York: Penguin.

Terdiman, D (2005). Hacking's a snap in Legoland (15 September 2005). *CNET News.com*.

Thomke, S (2002). *Innovation at 3M Corporation (A) and (B)*, Harvard Business Case Studies. Cambridge: Harvard Business School Publishing.

Tietz, R, J Füller and C Herstatt (2006). Signalling — an innovative approach to identify lead users in online communities. In *Customer Interaction and Customer Integration*, T Blecker and G Friedrich (eds.), pp. 453–467, Berlin: GITO-Verlag.

Tietz, R, C Herstatt, P Morrison and C Lüthje (2005). The process of user-innovation: A case study in a consumer goods setting. *International Journal of Product Development*, 2(4), 321–338.

Varian, H (2002). New chips can keep a tight rein on consumers (4 July 2002). *New York Times*.

von Hippel, E (1988). *The Sources of Innovation*. Oxford: Oxford University Press.

von Hippel, E (2005). *Democratizing Innovation*. Cambridge, MA: MIT Press.

von Hippel, E and G von Krogh (2006). Free revealing and the private-collective model for innovation incentives. *R&D Management*, 36(3), 295–306.

Wagstaff, J (2002). How to squeeze more juice from pricey printer cartridges (6 August 2002). *Wall Street Journal Online*.

Wellen, A and L Reiter (1998). Are hackers criminals? (12 May 1998). *TechTV Vault*.

Wiedhahn, K (2006). 20 Years Cerec. *International Journal of Computerized Dentistry*, 9(1), 1–4.

Zittrain, J (2005). In praise of uncertainty. *Harvard Business Review*, 83(5), 18–20.

Index

absorptive capacity, 105, 107, 176
actor degree centrality, 113
advanced manufacturing technology, 217
agency perspective, 169
all-terrain bikes, 19
antiprogram, 3
'anti' user innovation, 7
Apache, 24
application-specific integrated circuits (ASIC), 23
applied ethnography, 141
"attuning process", 174

barriers, 35
barriers to user innovation, 52
basketball, 25
biotechnology, 161
Bladerider International, 39
bloggers, 101
blogosphere, 96, 101, 115
boundary objects, 176
boundary spanners, 65
bounded rationality, 104, 108, 110
breakthrough, 6
breakthrough innovations, 44, 91
breakthrough product innovations, 147, 154, 155
bridging activities, 66
broadcasting, 93, 116
broader definition of innovation, 214
brokering, 67, 69, 70
business consultancies, 64

Canada — measuring user innovation in manufacturing, 217
category appraisal, 141
Cathedral and the Bazaar, 23
changing technologies and regulations, 18
chemical analysers, 15
co-design, 79
cognitive shortcomings, 108

commercialisation, 4
communities, 2, 20, 92, 95, 108
community of practitioners, 20
community-based innovation, 93
competence destroying, 132
competence enhancing, 132
complementary users, 176
computer games, 25
Computer-Aided Design (CAD), 22
computer-assisted telephony integration systems (CTI), 23
computer-controlled music instruments, 193
concept tests, 154
conceptualisation, 94, 114
configuration, 68
configurational technology, 165
configuring, 67, 70
conjoint analysis, 141
constructive technology assessment, 165
consultants, 66
consumer communities, 25
consumer goods, 19, 20
'consumer' is shaped, 57
consumer idealised design, 141
consumer involvement, 132
consumers, 163
content developers, 62
content service providers, 62
continuous, 132
conversation tracker, 101
coronary shunts, 17
crowd sourcing (develop), 141
cultural proximity, 177
cultural studies, 165
customer satisfaction, 49
customer-active paradigm, 15, 92
customer-driven innovation, 37
cybercafes, 68

Del.icio.us, 95
demand articulation, 164, 166, 173, 180

democratisation, 25
democratisation of innovation, 13, 35
Denmark — National programme for
   user-driven innovation, 216
design knowledge, 78
developers, 64
diffusion, 3
Digg, 95
digital products, 11, 13, 22, 25
digital tools, 13, 22
discontinuous, 132
discovery, 6
disruptive innovation, 132
domestication, 78, 82, 164
dominant design, 51
drug development pipelines, 161
Duchenne muscular dystrophy, 171
Dutch Nutrigenomics Consortium (DNC), 174
dynamics of user innovation, 46

ecological niches, 78
ecologies of intermediation, 60
ecology of intermediaries, 81
educational sector, 79
electron microscopy, 15
empathic design, 141
empirical data on medical instruments, 18
end users, 62, 64
enterprise resource planning systems, 79
"event history analysis" method, 170
evolutionary economics, 62, 165
exemplification projects, 176
experiential knowledge of users, 164
extreme needs, 112
extreme sports, 93
extrinsic motivation, 111

Facebook, 95
facilitating, 67, 70
feminist literature, 164
first-order learning, 166
flexibility of technology, 179
Flickr, 95
Fogarty catheter, 17
frame sharing, 164
free revealing, 2, 11, 20, 24, 27
free riders, 42
freely reveal, 216
functional fixedness, 95
fuzzy front end, 94, 114

Gartner Group, 71, 80
generification, 80
genomics, 161, 168
girl games, 66
government, 163

hacker communities, 216
hacker culture, 22
hackers, 2
hacking, 6
health-monitoring, 73
heterogeneity of the user population, 165
heterogeneity of user needs, 24
heterogeneity of user populations, 179
heterogeneous users, 168, 179, 182
'hidden innovation', 219
"high-bandwith" oral communication, 52
high expected benefit, 115
hijacking, 42, 50
historical context in user innovation, 4
homebrew software, 192, 197
homogeneous user populations, 165
homogeneous users, 182
horizontal innovation communities, 11
horizontal user networks, 192
hydrofoil technology, 38

idea-generation, 94
ideal consumer session, 139
ideation, 114
incremental, 132
incremental product innovations, 133
incremental products, 133
incremental return to tinkering, 48
incubation, 6
industry consultants, 64
industry-user fora, 68
informal communication, 113
informal intermediaries, 78
informal learning, 70
information acceleration, 141
Innobarometer, 221
innofusion, 79, 164
innofusion-domestication, 80
innovation barriers, 50
innovation communities, 191, 193
innovation cycles, 63
innovation in use, 59
Innovation Index, 220
innovation intermediaries, 58

Innovation Management, 1, 6
innovation metrics and indicators, 212
Innovation Nation White Paper, 219
innovation policy, 211
innovation studies, 1–6, 12–14, 27
innovation templates, 141
innovations, 132
innovative communities, 93
innovative users, 96
interactive learning, 166, 168, 179, 181, 182
intermediaries, 5, 21, 57, 63, 64, 66, 67, 70, 71, 73, 75, 80, 82, 170, 176, 180
intermediary, 6
intermediary institutions, 58
intermediary organisations, 59, 62, 72, 168
intermediary roles, 72
intermediary user organisation, 181, 182
intermediate and final users, 63
intermediate users, 62, 64, 80, 81
intermediation in innovation networks, 58
International Moth Class, 36, 38
International Pompe Association (IPA), 172
intransigent gatekeepers, 66

kitesurfing, 20
knowledge is tacit, 175
knowledge-brokering, 63

Last.fm, 95
latent consumer needs, 139
lateral thinking synectics, 141
lead user experiments, 51
lead users, 2, 3, 5, 11, 36, 92, 95, 96, 102, 111
lead-user characteristics, 104, 114
lead-user identification, 96
lead-user method, 91–94, 118, 141
lead-user theory, 102
lead-users, 78
leading-edge customers, 91
leading-edge users, 93, 101
learning by interaction, 164
learning by using, 164
learning economy, 60, 62
learning-by-doing, 61
learning-by-interacting, 61
learning-by-regulating, 61
learning-by-using, 16
learning-though-innovating, 61
linear innovation model, 164
linear model of innovation, 3, 7, 11, 211, 212

'linear' process model of innovation, 213
living in the future, 95
local experts, 65, 70, 81
"long and thin" intermediaries, 65

machine tools, 16
manifest needs, 134
manufacturer-active paradigm, 36
market failure, 215
market gap, 75
marketing and advertisement agencies, 78
mass customisation, 25, 93, 94
medical and scientific instruments, 11
medical instruments, 16–18, 26, 27, 216
medical practitioners, 19
medical professionals, 18
mountain biking, 19, 20, 40
MySpace, 95

nanotechnology, 161, 183
National Endowment for Science, Technology and the Arts (NESTA), 214, 220, 222
national innovation systems, 60
need-driven stimulus, 153
need-forecasting laboratory, 92
netnography, 93
network "bridgers", 60
networking, 93
new intermediaries, 82
new product development, 17, 26, 35
new sociology of markets, 57
Ning, 95
non-traditional intermediaries, 66
nuclear magnetic resonance (NMR), 15
nutrigenomics, 168

OECD, 222
offline communities, 109
offline life, 96
offline world, 95
online communities, 95, 102, 103, 109
Online Public Access (OPAC), 23
online world, 95
open innovation, 57, 93
open-source, 2, 23, 82, 216
open source movement, 194
open source software, 19, 27, 52, 107
OSS, 23
outdoor sporting equipment, 13
outdoor sporting goods, 11

outlaw, 27
outlaw communities, 7, 192, 194, 195
outlaw community innovations, 6
outlaw innovation, 191, 197, 205, 207
outlaw users, 6

patent, 213
patent protection, 50
performance-enhancing modifications, 39
permalinks, 101
pharmaceutical industry, 162
pharmacogenomics, 183
pirated games, 192
pompe disease, 171
proactive consumer involvement, 154
probe and learning, 141
producers, 163
product-driven stimulus, 153
product-related knowledge, 105, 109, 110, 114, 117, 118
professional intermediaries, 76
Project SAPPHO, 14
Propellerhead, 25
proxy users, 62, 76
Pruitt-Inahara Carotid shunt, 17
pyramiding, 93, 115, 117

radical, 132
radical innovation, 5, 133
radical product innovation, 131, 133, 139, 154
radical user ideas, 52
re-configuration, 69
re-innovation, 3
re-inventions, 60
reactive research, 149
really new products, 133
regulatory barriers, 27
regulatory hurdles, 161
regulatory proximity, 177
Research and Development (R&D) metrics, 213
research institutes, 163
Research Policy, 14
rodeo kayaking, 20, 21
rodeo kayaks, 13
rollerblading, 19
R&D, 13, 151
R&D specialists, 139

sailing, 35
Science and Technology Studies, 1, 3, 4, 6

science push, 13
Science, Technology and Innovation (STI) indicators, 212
science-based innovation, 163, 164
scientific instruments, 12, 14–16
screening, 115, 117
script, 3
second-degree relationships, 109
second-order learning, 166
selection of consumers, 151
semiotic studies, 165
"short and fat" intermediaries, 65
skateboarding, 19
snowboarding, 19, 20
social construction of technology, 165
social identity, 193
social learning, 62, 67
Social learning in Multimedia, 74
Social Learning in Technological Innovation (SLTI), 5, 58
social networking, 95
social perception, 105, 107
social psychology, 99, 107
social ties, 109
social welfare gains, 216
software, 216
software hackers, 194
spectroscopic instruments, 16
speed of adoption, 103
spinal cages, 17
sporting equipment, 26, 216
sporting goods, 20
sports equipment, 4, 35
statistical software, 193
"sticky" information, 110
sticky knowledge, 11, 23, 65
structure perspective, 169
'super-innovators', 222
supplier representatives, 64
sustainable innovation, 132
symbiotic relationship from which both users and manufacturers, 191

tacit knowledge, 177
tailors, 65, 70
technological maturity, 47
technologically really new, 6, 154
technologically really new product innovations, 142, 151
technology complexity, 46
Technology Experiment, 77

technology mediators, 65
technology-need matrix, 133
The EU — towards measuring user innovation across 27 countries, 221
The Netherlands — measuring user innovation in SMEs, 218
The UK — Measuring user innovation in firms and individual consumers, 219
tinkerers, 40
toolkit, 3, 11, 22, 24, 25, 27, 93, 94, 141, 216
tough customers, 2
trade associations, 71
transactive memory, 107
trend-break really new, 6, 155
trend-break really new product innovations, 144, 152
trial-and-error, 61, 108
Twitter, 95

universities, 163
usability consultants, 78, 82
use experience, 106–108, 110, 118
use-side intermediaries, 65
user communities, 4, 27, 192
user communities of a proprietary software, 193
user customisation, 15
user dissatisfaction, 103
user entrepreneurship, 17
user experience, 105
user expertise, 104
user groups, 71
user initiated innovation, 59

user innovation communities, 78
user innovation indicators, 222
user interaction, 131
user investment, 102, 117
user involvement, 162
user needs, 14
user networks, 35
user-acceptance, 91
user-as-innovator, 12
user-centred design, 76
user-centric innovation, 92
user-communities, 57
user-designer community, 78
user-designer relationship, 81
user-end intermediaries, 80
user-involvement, 80

video games, 193, 196
"visionary skills" of consumers, 155
visioning/back casting, 141
voice-of-the-customer, 91, 163
VSN (Vereniging Spierziekten Nederland), 171, 172

waves of technological change, 28
weak and strong ties, 105
Web 2.0, 5, 91, 93, 95, 96, 103, 106, 113, 117
weblogs, 95, 101
Wikipedia, 95
windsurfing, 19, 20
wisdom of crowds, 91

YouTube, 95